Space Science, Exploration and Policies

nova
science publishers
www.novapublishers.com

Space Science, Exploration and Policies

Conceptual Features of Einstein's Theory of General Relativity Based on the Philosophy of Science
Jun-Young Oh
2022. ISBN: 979-8-88697-107-1 (Softcover)
2022. ISBN: 979-8-88697-140-8 (eBook)

Time-Out of Time: A Postscript to Nuclear Time Travel
Bernd Schmeikal, PhD
2021. ISBN: 978-1-68507-320-6 (Softcover)
2021. ISBN: 978-1-68507-465-4 (eBook)

Solar System: Structure, Formation and Exploration
Matteo de Rossi (Editor)
2012. ISBN: 978-1-62100-057-0 (Hardcover)
2012. ISBN: 978-1-62100-081-5 (eBook)

Ultra High Energy Cosmic Rays: A New Tool for Astrophysics Research
Anna Uryson
2011. ISBN: 978-1-61668-847-9 (Softcover)
2011. ISBN: 978-1-61728-258-4 (eBook)

An Introduction to Magentometers
Thembinkosi Donald Malevu, PhD, Zolile Mtumela, PhD (Editors)
2020. ISBN: 978-1-53618-756-4 (Hardcover)
2020. ISBN: 978-1-53618-847-9 (eBook)

More information about this series can be found at
https://novapublishers.com/product-category/series/space-science-exploration-and-policies/

Kenath Arun
Editor

A Guide to Black Holes

nova
science publishers

www.novapublishers.com

https://doi.org/10.52305/JOLT6397

Library of Congress Cataloging-in-Publication Data

ISBN: 979-8-88697-163-7

Published by Nova Science Publishers, Inc. † New York

Contents

Preface

The idea of a black hole goes back to Laplace or maybe even earlier to Englishman John Michell, referring to these bodies as 'dark stars', way back in 1783. Laplace argued that the largest objects in the universe must become invisible and would hence be dark. Laplace argued that if an object – of the same average density as the Sun – were about 300 times larger than the Sun, the escape velocity for the object would exceed that of light. When Einstein proposed his general theory of relativity in the early 1900s, Schwarzschild solved the equation (1916) for a spherical star and the solution implied that below a radius (referred to as the Schwarzschild radius), all light and other radiation are trapped inside the star, and it becomes what is called a black hole. Coincidentally the expression for the Schwarzschild radius agrees with that arrived by Laplace.

The term black hole became coined for popular universal usage apparently in an article by John Wheeler in Physics Today, 1971, "Introducing the Black Hole". The formation of such a trapped surface as a star collapse beyond the Schwarzschild radius ensures that no matter or radiation can escape from the object once this radius of the trapped surface is reached. This radius depends only on mass and is linearly proportional to it. For a star of the Sun's mass, this critical Schwarzschild radius is about 3 kilometres. An object like the Earth has to shrink to about one centimetre (the size of a marble) for it to become a black hole.

With better detection mechanisms, such as the images from the Event Horizon Telescope, the universe of black holes is opening up for astronomical observations and discoveries over all wavelengths. Possible 'black hole' evidence first came with the discovery of the first quasars. Later, evidence for black holes in the Milky Way was found in Cygnus X-1, an X-ray emitting black hole candidate. The theoretical framework, in the meantime, was developed by Roger Penrose. This, along with the observations of the supermassive black hole (SMBH) at the centre of our Milky Way galaxy, was recognised with the awarding of the 2020 Nobel Prize in physics to Roger Penrose, Andrea Ghez and Reinhard Genzel.

In this edited book, we have a wide range of topics covering various aspects of black holes and their dynamics. *Invoking dark matter in the formation of SMBH in the very early Universe: Possible solution to a continuing conundrum* looks at the formation mechanism of supermassive black holes. The two possible mechanisms discussed are the mergers of intermediate-mass black holes to form the SMBH, and the role of dark matter in providing the seed for the formation of SMBH. Also discussed is the possible connection between the gravitational binding energies of large galaxies and the masses of their central black holes.

Quantum black holes and QFT applicability boundary discusses the results from the use of the concept of measurability in Quantum Field Theory (QFT). The results are applied to study the black hole thermodynamics, in particular the thermodynamics of quantum black holes. In the considered case, these are black holes with Planck values of horizon radius and mass.

Thermodynamics of black hole and thermal fluctuations discusses basic concepts of black hole thermodynamics based on the Hawking assumption as to what happens to the thermodynamics of black holes when their sizes are reduced, and looks at how the small black holes get influenced by small statistical fluctuations around an equilibrium. This is then extended to the black hole solution for the charged quasitopological gravity (which includes Gauss-Bonnet term and curvature-cubed interactions). The correction in thermodynamics for such black holes when fluctuations around equilibrium are also discussed.

Motion of Spinning Particles around Black Holes presents a computational code that integrates the differential equations describing the motion of a spinning particle in the equatorial plane of a general spherically symmetric space-time arising from general relativity. This code is open source, so that users may extend its application not only to the Schwarzschild metric, but to other spherically symmetric backgrounds. It gives a visual representation of the trajectory of the spinning particle in the plane of motion and the code can be easily modified to include other physical effects and even other gravitational theories.

Testing a microscopic model for black holes deduced from maximum force combines the two limits that imply Einstein's field equations of general relativity, with the quantum of action and an idea by Dirac. This leads to a model for the microscopic degrees that make up freedom of space, particles and horizons. This model is in agreement with all aspects of black hole physics. In addition, the same tangled microscopic degrees of freedom lead to the standard model of particle physics.

From Bohr to Schrodinger: Two-particle system approach to black hole quantum gravity develops a new approach to black hole quantum gravity. The Schwarzschild BH (SBH) surfaces are treated as a quantum non-singular two-particle system where the two components, the nucleus and the electron, are subject to a mutual quantum gravitational interaction. The semi-classical Bohr-like approach to BH quantum physics is also discussed. The chapter goes on to show that the SBH emerges as the gravitational analogous of the hydrogen atom.

Black Hole Menagerie, Charged/Dyonic BHs and Radiation from Interacting Dyonic BH Pairs looks at the exposition of basic physics related to black holes and their event horizon, providing the physical reasons why SMBHs are highly efficient engines that convert matter into radiation and explaining the bizarre behaviour of quasars and blazars. Challenges associated with SMBH formation, production of primordial black holes in the early universe as well as their quantum aspects like Hawking radiation and the ensuing information loss paradox are described. Magnetic monopoles, dyons (hypothetical particles bearing both electric as well as magnetic charges) and dyonic black holes are addressed. The physical quantities of dyonic black holes are computed so that their characteristic signatures can be identified.

Chapter 1

Invoking Dark Matter in the Formation of SMBH in the Very Early Universe: Possible Solution to a Continuing Conundrum

Arun Kenath[1] **and C. Sivaram**[2*]
[1]CHRIST (Deemed to be University), Bangalore, India
[2]Indian Institute of Astrophysics, Bangalore, India

Abstract

Most of the large galaxies host a supermassive black hole, but a continuing astronomical conundrum is the presence of such supermassive black holes that seemed to have formed hardly a little more than half a billion years after the expansion of the Universe began. Here we explore the possibility that clumps of dark matter could have collapsed (before first stars formed), forming seed primeval black holes. We also look at the possibility of intermediate mass black hole dynamics and its merger as a possible mechanism for formation of supermassive black holes. Another intriguing question is whether galaxies formed first or their central black holes? We look at a possible connection between the gravitational binding energies of large galaxies and the masses of their central black holes. We also give some interesting properties of extremal black holes (of any mass).

Keywords: supermassive black holes, intermediate mass black holes, extremal black holes, dark matter

*Corresponding Author's Email: kenath.arun@christuniversity.in

In: A Guide to Black Holes
Editor: Kenath Arun
ISBN: 979-8-88697-163-7

1. Introduction

Supermassive black holes (SMBH) are black holes with a mass from about 0.1 million to over a 1 million solar mass ($M_\odot$). Most of the large galaxies host an SMBH at its core, but one of the most pressing astronomical conundrum is the presence of such SMBH that seemed to have formed hardly a little more than half a billion years after the expansion of the Universe began. The time scale involved in the growth of such black holes even at Eddington accretion rates is $\sim 6 \times 10^8$ years.The seed SMBH should have been quite massive. How then did that form? Scenarios involving collapse of a large number of early massive stars forming and rapidly evolving in a compact region of an emerging primeval galaxy have been suggested, among other hypotheses. For a description of early work see for e.g., [1].

One possible scenario is that clumps of dark matter could have collapsed (before first stars formed), forming such seed primeval black holes [2]. As dark matter does not couple to radiation, there is no constraint in Eddington accretion rates as in the case of baryonic matter, i.e., no radiation pressure to stop dark matter in the early universe to collapse directly. Massive stars forming from the baryonic matter accreting onto such seed dark matter black holes, could form such SMBHs on a time scale of half a billion years, as observed.

As galaxies were closer at the earlier epochs, galaxy mergers would also have led to merger of SMBHs in their cores producing the most massive SMBHs. Merger of intermediate mass black holes (IMBHs) is not a likely scenario [3].

2. Possible Scenarios for Formation of SMBHs

2.1. Merger of IMBH

Intermediate mass black holes (IMBH) are those black holes having mass between that of stellar black holes and supermassive black holes, i.e., in the range of $500 - 10^4 M_\odot$. Zwart, et al [4] propound that dense star densities near galactic centres can lead to runaway stellar mergers thus efficiently producing IMBH's. Indeed, they suggest that the innermost ten parsecs of our galaxy can contain about 50 IMBH's, each of about a thousand solar masses. These can sink towards the core as they interact with the stars. Every time an IMBH has a stellar encounter, the stars' velocity is boosted. This reduces the potential energy making the IMBH sink to the galactic core. These IMBH's can thus ultimately merge with the galaxy's supermassive black hole.

However if very near the galaxy SBH there are not enough stars, the IMBH may stop falling inward, halting at about 0.01 light years from the centre. However if there are several IMBH's near the galactic core, they can interact with one another and once every million years an IMBH can merge with the SBH.

It has been suggested that presence of several IMBH's near the galactic centre might explain why there are so many clusters of young stars (like S2, the B type star) in the galactic core, where tidal forces should rip apart the gas clouds from which the stars form. These stars could have formed much further away and the IMBH's could have shepherded them inwards, much like the thin rings of Uranus or Saturn that are kept in place by being shepherded' by tiny satellites.

This IMBH can merge with the stars in the surrounding volume, ($10^4 M_\odot/pc^3$), to give a supermassive black hole in the time scale given by,

$$t = \frac{v^3}{nG^2M^2 lnN} \approx 10^8 years \tag{1}$$

This is the time scale for relaxing in a cluster of N objects of average mass M moving with a mean velocity of v If $\sigma$is the average cross-section for interaction, $\sigma \approx (\frac{GM}{v})^2$, where $(\frac{GM}{v})$ gives the 'radius' of the sphere of influence around a given individual object and the mean free path is $1/n\sigma$, n is the number density of objects. Then the 'collision' or 'interaction' time scale is given by:

$$t \approx \frac{1}{n\sigma v} = \frac{v^3}{nG^2M^2} \tag{2}$$

The lnN term comes from many body effects (i.e., $\int \frac{dN}{N}$, the so called Coulomb logarithm in plasma physics). Note the sharp v^3 dependence as well as $\frac{1}{nM^2}$.

Other possible ways in which SMBH can form involving IMBH are by the merger of two or more IMBH, or by accretion of matter by the IMBH in systems such as AGNs and quasars. Moving masses like black holes produce gravitational waves in the fabric of space time. A more massive moving object will produce more powerful waves, and objects that move very quickly will produce more waves over a certain time period.

Gravitational waves are usually produced in an interaction between two or more compact masses. Such interactions include the binary orbit of two black holes orbiting each other. As the black holes orbit each other, they send out waves of gravitational radiation that reaches the Earth, however, once the waves do get to the Earth, they are extremely weak. This is because gravitational waves decrease in strength as they move away from the source. Even though they are weak, the waves can travel unobstructed within the fabric of space-time. From Kepler's third law, the period is related to the separation by,

$$P^2 = \frac{4\pi^2}{G(M_1 + M_2)} R^3 \tag{3}$$

Knowing the period, we can determine the orbital velocity from, $vP = 2\pi R$ Power lost by gravitational waves (quadrupole formula for gravitational waves) is given by (for objects of mass M with separation of R):

$$\dot{E}_{GW} = \frac{32}{5}\frac{G}{c^5} M^2 R^4 \omega^6 \tag{4}$$

This emission results in the objects coming closer. The merger time is obtained by integrating the rate of binding energy change:

$$\frac{GM^2}{R^2}\frac{dR}{dt} = \frac{32}{5}\frac{G}{c^5} M^2 R^4 \omega^6 \tag{5}$$

This implies a merger time for two objects separated by an initial distance R_0 of:

$$\tau_{mer} = K R_0^4 \tag{6}$$

With the constant, $K = \frac{5}{256}\frac{c^5}{G^3}\frac{1}{M^3}$for equal mass objects.So a larger initial separation would imply a longer merger time. So for the merger time to be $\approx 10^9$years or less, the initial separation can be precisely calculated.

For two IMBHs of mass $M = 10^4 \mathrm{M}_\odot$each, to merge in $\tau_{mer} = 10^9$ years, the distance of separation should be of the order of,

$$R = \frac{GM^2}{\tau_{mer} \dot{E}_{GW}} \tag{7}$$

where,

$$\dot{E}_{GW} = \frac{128 v^{10}}{5Gc^5}, v = \frac{2\pi R}{P}, P = \sqrt{\frac{4\pi^2 R^3}{G(M_1 + M_2)}} \Rightarrow v = \sqrt{\frac{GM}{R}} \tag{8}$$

And the energy loss due to the gravitational waves emission is given by:

$$\dot{E}_{GW} = \frac{128}{5Gc^5}\left(\frac{GM}{R}\right)^5 \tag{9}$$

Therefore the distance of separation is given by:

$$R = \left(8 \times 10^2 \frac{(GM)^3 \tau_{mer}}{c^5}\right)^{\frac{1}{4}} \approx 2 \times 10^{12} m \tag{10}$$

The corresponding orbital frequency is given by,

$$f = \frac{v}{2\pi R} \tag{11}$$

The orbital velocity is of the order of,

$$v = \sqrt{\frac{GM}{R}} \approx 10^6 ms^{-1} \tag{12}$$

This implies that the frequency is,$f \approx 10^{-7} Hz$, and hence the period will be given by, $P \approx 10^7 s$ ($\sim$ a year).

For two IMBH with separation of about 10^4 parsec, the time taken to merge is of the order of Hubble time. For such a model to produce a SMBH, we need about 10^3 IMBH merging together. Hence the model with the IMBH merging with the surrounding stars, gives a much more efficient way of generating a supermassive black hole.

In the case of accretion of matter by IMBH to form SMBH, the increase in mass is exponential with time [2]:

$$M = M_o e^{kt} \tag{13}$$

where, $k^{-1} = t \approx 6 \times 10^8$ years is the characteristic time required for the mass to increase e - folds, with the accreting disk emitting at maximum luminosity. For the IMBH to accrete enough matter to become SMBH of mass say $10^8 M_\odot$, we have:

$$e^{kt} = \frac{M}{M_0} = 10^4 \tag{14}$$

The corresponding time scale is of the order of

$$t \approx 5 \times 10^9 years \tag{15}$$

This implies that the mass will have to increase $\sim e^{10}$ folds by accretion for the IMBH to become an SMBH. Hence even this model does not provide an efficient way of formation of SMBH from an IMBH.

IMBHs would have formed in the early universe. Owing to low metal content, the earliest stars would have been very massive, a few hundred solar masses [5]. Such stars would end up in a pair-instability supernova (around oxygen-neon burning temperature of 2 billion degrees) and would collapse into a black hole (if their mass exceeds 250 solar mass).

Oxygen-neon burning occurs at $\approx 2 \times 10^9 K$. At this stage, temperatures are high enough for the electron-positron pair production processes ($\gamma \rightarrow e^+ + e^-$). At $\approx 6 \times 10^9 K$, it is maximal. Now if the nature of the black body radiation changes from photons to electron-positron pairs (which are fermions) the energy density is no longer aT^4, but $\frac{7}{8}aT^4$. So once the pairs are forming, the radiation pressure drops by $\frac{1}{8}aT^4$. So there is less support against gravitational collapse or gravitational contraction. The balance between the two (or the 'stability condition') can be expressed as:

$$\frac{1}{8}aT^4 \geq \frac{GM^2}{R^4} \tag{16}$$

Now, $T \approx 2 \times 10^9 K$, $R \approx 2 \times 10^8 m$ (appropriate density for oxygen-neon burning), this implies a limiting mass M of 250 solar mass, beyond which the gravitational energy density is higher. This result in the context of population III supermassive stars is discussed by Heger and Woosley [6].

Signatures of such explosions of supermassive stars at $z \approx 10$ (when the universe was only half a billion years old, and eleven times smaller) could be sought with future space telescopes [7] [8].

Observations of galaxy core show correlation between black hole masses and the spheroidal (bulge) component of the host galaxy. Thus the primordial low mass galaxies (blue galaxies) would host low mass central black holes, which just correspond to the IMBH mass (about 10^5 10^6 the galaxy mass). There is a well-known relation between black hole mass and spheroidal bulge component mass of the host galaxy [9]. Typically, the black hole mass is 10^3 the bulge mass. Primordial galaxies (so called blue galaxies as seen for example in Hubble deep field) have a lower mass $\sim 10^8$ solar mass. This would give a central black hole mass of $\sim 10^4$ solar mass, corresponding to a typical IMBH. Merger of these primeval galaxies would lead to larger galaxies and cluster and the IMBH's would also merge and sink to the core forming an SMBH (like for instance $M87$ has a $6 \times 10^9 \mathrm{M}_\odot$ black hole).

2.2. DM Admixture

Recently astronomers combining data from the WISE telescope and ground based surveys involving also the Magellan telescopes in Chile have uncovered evidence for a 7 supermassive black hole having a mass of 800 million solar mass [10] [11] at a redshift of about 7.6. This would imply that the black hole grew to this mass only 690 million years after the universe started expanding, 13.8 billion years ago. This SMBH sits at the centre of a galaxy, i.e., a quasar. How can a black hole have such a large mass so soon in the history of the universe? This is regarded as a puzzle. Formation of such SMBH's is a conundrum [12].

Here we envisage a possible scenario. First we observe that the time period of around 600 million years, curiously corresponds to the so-called Salpeter time scale given by [13]

$$t_{sal} = \frac{c\sigma_T}{4\pi G M_p} = 6 \times 10^8 years \tag{17}$$

Here, m_p is the proton mass and σ_T is the Thompson cross section given by:

$$\sigma_T = \frac{8\pi}{3}\left(\frac{e^2}{m_e c^2}\right)^2 \approx 10^{-24} cm^2 \tag{18}$$

m_e, e being the electron mass and charge.

This time scale also corresponds to the time taken for the black hole to grow to 'e' times its initial mass, i.e., e-folding growth time. As is well known, this arises for instance, when considering the black hole (of mass M_{BH} emitting at the maximal Eddington luminosity (while accreting ambient matter), given by [14]:

$$L_{Edd} = \frac{4\pi GMBHm_p c}{\sigma_T} \tag{19}$$

The accretion luminosity is the gravitational energy released per unit time, i.e.,

$$L_{grav} = \frac{GM_{BH}\dot{m}}{R} \tag{20}$$

with $\dot{m}$ being the mass accretion rate. Putting $\frac{2GM_{BH}}{c^2}$ and equating (19) and (20), gives the time scale for the black hole mass increase as:

$$\frac{dM_{BH}}{M_{BH}} = \frac{4\pi G m_p}{c\sigma_T} dt \tag{21}$$

which gives: [15] [16]

$$M_{BH} = M_{BH(0)} exp \left(\frac{4\pi G m_p}{c\sigma_T} t \right) \tag{22}$$

i.e. e-fold increase time scale, given by equation (17), of just 6×10^8 years [2].

To grow to $8 \times 10^8 M_\odot$, in 6×10^8 years, the initial mass $M_{BH(0)}$ must be $3 \times 10^8 M_\odot$. How could $M_{BH(0)}$ be this large? A possibility is the collapse of a cloud of dominantly dark matter (DM) particles, which could have started collapsing even at redshift $z = 50$ or 100, as unlike baryonic (or atomic) matter, DM would not radiate (and lose energy) not being coupled to electromagnetic radiation. The ambient DM density [17] at $z = 100$, is $\rho_{DM} \approx 10^{-23} g/cc$, which gives a free fall (hydrodynamic) collapse time scale:

$$t = \frac{1}{\sqrt{G\rho_{DM}}} \approx 3 \times 10^7 years < 10^8 years \tag{23}$$

A cloud of such DM dominated matter (with a significantly smaller baryon fraction) having a radius of 200pc, would have a mass, $M_{DM} \approx 3 \times 10^8 M_\odot$. This would have collapsed to a black hole on a time scale of $\sim 10^8$ years.

This seed SMBH would now grow by accumulating the surrounding gas (which collapses on to the black hole), to $8 \times 10^8 M_\odot$ according to the equations (19) - (22), on a time scale of 6×10^8 years. So the total time scale for forming a SMBH is $\sim 7 \times 10^8$ years, which corresponds to a redshift of $z \sim 7.5$. The black hole accreting the (predominantly hydrogen) gas would radiate at L_{Edd}. An accretion rate of one solar mass a year would enable it to grow to the observed SMBH mass. It would radiate at $\sim 10^{40} watt$, a typical quasar luminosity. Smaller DM clouds, with higher density, at even higher redshifts would also have collapsed to form such seed black holes. (One cannot distinguish a seed DM black hole from that of a black hole formed by any other kind of matter, including baryonic).

Also the collapsing hot gas could form clusters of massive stars (like the Arches cluster in the centre of the Milky Way) [18] [19]. These short lived clusters of massive stars (lifetimes of a few million years) could explode as supernovae, the relaxation time of the clusters is again $\sim 10^8$ years, and these massive matter will also collapse back into the black hole. The above scenario suggests that formation of such SMBH's so early in the universe is consistent with our present understanding of the phenomena involved.

3. Central Black Holes and Gravitational Binding Energy of Large Galaxies

It is by now well established that most large galaxies (spiral, elliptical, etc.) host a supermassive black hole (SMBH) in their centre [20]. Again AGN's, quasars, etc. are powered by the gravitational energy of matter accreting on to these central SMBH [21]. These black hole masses are typically of several millions of solar masses and can be as large as a few billion solar masses, like in the case of $M87$. The origin of these supermassive black holes at the galactic cores is still an enigma. Again the black hole masses have been related to the galactic bulge mass, suggesting a common origin [22] [16].

Various studies have pointed to the correlation between the masses of the SMBHs at the centre of galaxies and the different characteristics of their host galaxies[23], such as the spheroid luminosity and spheroid mass [24], the spheroid velocity dispersion [25], and the kinetic energy of random motions [26]. Kormendy relation connects the estimated mass of the black hole to the K-band luminosity and the velocity dispersion of the bulge of the host galaxy measured outside the BH sphere of influence [27].

But the underlying physics of these various observations is not yet fully understood. It is thought that [28] feedback processes are responsible for such correlations. Also there is no understanding as to whether these relations evolve with evolution of the galaxy (and the central BH) or set at the time of their formation [29]; [30]. Again, many other large stellar conglomerations like globular clusters or dwarf galaxies also do not seem to host massive black holes in their core regions.

Here we propose a possible connection between the gravitational binding energies of large galaxies etc. and the masses of their central black holes[31] . The idea is similar to what happens in core collapse of massive stars. The gravitational binding energy released in stellar collapse is carried away by neutrinos mainly (like in the case of SN1987A). The total energy carried away is just the gravitational binding energy of the remnant neutron star [32].

In the case of SN1987A, the estimated $3 \times 10^{53} ergs$ carried away by neutrinos, just corresponds to the binding energy of a $1.4 M_{\odot}$ neutron star [33]. More massive stars would release more binding energy and this would correspond to the formation of a black hole. The gravitational binding energy released in the formation of a black hole would correspond to its rest energy, i.e., to the mass of the black hole. (For the same reason, a supernova would not give rise to a white dwarf as a remnant, as the binding energy of the white dwarf is two or more orders smaller than the energy released in the explosion).

By analogy, we argue that whatever processes led to the formation of the black holes at the galaxy cores were connected with the formation of the galaxy and should thus be related to the gravitational binding energy of the final structure which forms [34]. We note that in case of galactic structures, the binding energy of the parent galaxy is comparable to that of the black hole hosted by the galaxy. The binding energy of a typical galaxy like the Milky Way is given by:

$$(BE)_{spiral} = \frac{GM^2_{spiral}}{R_{spiral}} \approx 10^{61} ergs \tag{24}$$

$(M_{spiral} \approx 10^{12} M_{\odot}, R_{spiral} \approx 10kpc)$
The galaxy harbours a $\sim 10^6 M_{\odot}$ black hole. The energy (mass) of the black hole is:

$$(BE)_{BH} = M_{BH}c^2 = \frac{GM^2_{BH}}{R_{BH}} \approx 10^{61} ergs \tag{25}$$

The equality of equations(24) and (25) is suggestive of what happens in the release of gravitational binding energy in stellar collapses leading to the formation of a remnant compact object.

Again for large elliptical galaxies, mass is few orders higher, that is: $M_{elliptical} \approx 10^{14} M_{\odot}$, so the binding energy is:

$$(BE_{elliptical}) \approx 10^{63} ergs \tag{26}$$

So galaxies like $M87$ harbour a billion solar mass black hole, whose binding energy:

$$(BE_{BH}) \approx 10^{63} ergs \tag{27}$$

Therefore we have in the case of the galaxies; the binding energy of the galaxy being comparable to the black hole energy:

$$M_{BH}c^2 = \frac{GM^2_{galaxy}}{R} \tag{28}$$

Hence the fraction of black hole mass to the galactic mass is given by:

$$\frac{M_{BH}}{M_{galaxy}} = \frac{GM^2_{galaxy}}{Rc^2} \approx 10^{-5} - 10^{-6} \tag{29}$$

This implies that smaller galaxies of $10^8 M_\odot$, could harbour black holes of $100 M_\odot$.
This is in agreement with the observations of host galaxy mass and its central BH mass as can be seen from table 1[34] [35]

Table 1. Correlation between SMBH mass (M_{BH})and mass of its host galaxy (M_{galaxy})

Galaxy system	$M_{BH}(M_\odot)$	$M_{galaxy}(M_\odot)$	$\frac{M_{BH}}{M_{galaxy}}$
SDSS J0029 - 0055	2.65×10^8	1.04×10^{13}	2.5×10^5
SDSS J0252 + 0039	6.48×10^7	1.27×10^{13}	5.1×10^6
SDSS J0330 - 0020	1.88×10^8	1.51×10^{13}	1.2×10^5
SDSS J0728 + 3835	1.90×10^8	1.73×10^{13}	1.1×10^5
SDSS J0737 + 3216	1.28×10^9	2.41×10^{13}	5.3×10^5
SDSS J0822 + 2652	4.67×10^8	2.00×10^{13}	2.3×10^5
SDSS J0955 + 0101	1.14×10^8	1.21×10^{13}	9.4×10^6
SDSS J0959 + 0410	2.25×10^8	1.08×10^{13}	2.1×10^5
SDSS J1142 + 1001	2.81×10^8	1.68×10^{13}	1.7×10^5
SDSS J1403 + 0006	1.74×10^8	1.18×10^{13}	1.5×10^5
SDSS J1451 - 0239	2.21×10^8	1.17×10^{13}	1.9×10^5
SDSS J1538 + 5817	1.06×10^8	1.17×10^{13}	9.1×10^6
SDSS J2238 - 0754	1.47×10^8	1.44×10^{13}	1.0×10^5
SDSS J2303 + 1422	3.83×10^8	2.57×10^{13}	1.5×10^5
SDSS J2341 + 0000	1.44×10^8	1.87×10^{13}	7.7×10^6

Result from equation(29) could also provide a reason as to why globular clusters do not host a black hole. The binding energy of the cluster is:

$$(BE)_{cluster} = \frac{GM^2_{cluster}}{R_{cluster}} \approx 10^{51} ergs \tag{30}$$

which is lesser than the binding energy of even neutron stars.

In the case of some globular clusters like the Mayall II (G1) in the Andromeda galaxy, there is evidence of a $2 \times 10^4 M_\odot$ black hole. There are also studies [36] that have suggested that the globular cluster G1 could in fact be the core of a dwarf elliptical galaxy. The BH in G1 could have been formed at an earlier epoch when the cluster was more closely packed or due to merger of few clusters. X-ray and radio emissions from Mayall II appear to be consistent with an intermediate-mass black hole. The energy of the $10^3 M\odot$ black hole is: $E = M_{BH}c^2 \approx 10^{57}$ ergs. In order for the cluster to have a binding energy of this order, its size should be:

$$(R)_{cluster} = \frac{GM_{cluster}^2}{M_{BH}c^2} \approx 10^{-4} pc \tag{31}$$

where, $M_{cluster} \sim 10^6 M_\odot$
As the potential decreases, the kinetic energy of the cluster increases, hence spreading it out over time.

One of the possible models for the formation of an IMBH is the collapse of a cluster of stars [37]. The collapsed core can accrete matter ejected during the formation. If a collection of a thousand 10 solar mass stars in a volume of a parsec cube collapses, ejecting 30% of its mass, then the total mass of the ambient gas,nm_p, is $\approx 6 \times 10^{33}$kg. If about 30% of the mass of each star is ejected, then the accretion rate given by, $\dot{m} = \frac{6\pi nm_p G^2 M^2}{c^3}$, is $\approx 4 \times 10^7$kg/s. For a constant density the dynamical friction [38], is given as:

$$F_d = C\frac{G^2M^2\rho}{v_m^2} \tag{32}$$

where, G is the gravitational constant, M is the mass of the moving object, ρ is the density, and v_m is the velocity of the object in the frame in which the surrounding matter was initially at rest. C is not a constant but depends on how v_m compares to the velocity dispersion of the surrounding matter.

The dynamical friction comes into effect in the evolution of the cluster due to the interaction between the ambient gas and dust and the central IMBH. The initial black hole is subjected to a gravitational force due to the star cluster, which is given by:

$$\nabla^2\phi(r) = 4\pi G\rho(r) = 4\pi Gmn(r) \tag{33}$$

where, $n(r)$ the density distribution of stars and m is the typical mass of the star. For a constant density, $\rho(r) = \rho_0$, the potential is:

$$\phi(r) = -2\pi G\rho_0 \left(R^2 - \frac{1}{3}r^2\right) \tag{34}$$

where R is the radius of cluster.
And the gravitational force on the black hole is given by:

$$F_g = -M_{BH}\nabla\phi(r) = -\frac{4}{3}\pi G\rho_0 M_{BH} r \tag{35}$$

(The particle inside a homogenous gravitational system performs simple harmonic motion.) And the equation of motion of the black hole in the star cluster is given by:

$$M_{BH}\frac{d^2r}{dt^2} + kr + \gamma\frac{dr}{dt} = 0 \tag{36}$$

where, $k = \frac{4}{3}\pi G\rho_0 M_{BH}$ and $\gamma = \frac{-3\pi G^2 mn(r)M_{BH}^2}{\left(\sqrt{2}\sigma\right)^3}$, σ is the velocity dispersion of stars in the cluster. The black hole undergoes a damped oscillation in the star cluster, with a damping time corresponding to $t_{damp} = \frac{M_{BH}}{\gamma}$

In the case of the system $M82$, which harbours an IMBH of mass in the range of $500M_\odot$ [39], the black hole is not found at the centre, but displaced by about one kilo-parsec from the centre. The frequency corresponding to the oscillation of the BH, ω, is $\sqrt{k/M_{BH}} \approx 10^{-13}s^{-1}$

For the $M82$ system, the number density of the stars in the cluster, $\approx 10^4/pc^3$, with a typical mass of $1M_\odot$ having a velocity dispersion, $\sigma \approx 30km/s$. We get the damping time for the system as, $t_{damp} = 4 \times 10^{12}s$, which is of the order of 10^5 years. With the effects of dynamical friction, the relaxation time for the system to form the IMBH is $t_{form} \approx 10^7$years. For a denser core, i.e., about $\approx 10^3/(0.01pc)^3$, the time taken to form the IMBH will be $\approx 10^6$ years[3].

Similarly, certain dwarf galaxies could also harbour $10^3M_\odot$ black hole, even though their binding energy (with visible mass) suggests that they shouldn't. The binding energy of the dwarf galaxy is:

$$(BE)_{dwarf} = \frac{GM_{dwarf}^2}{R_{dwarf}} \approx 10^{55}ergs \tag{37}$$

And the energy of the $10^3M_\odot$ black hole is of the order of 10^57 ergs. This would suggest that there is a substantial amount of dark matter to make up for the binding energy of the dwarf galaxy, i.e., [40]; [41]:

$$\frac{M}{L} = 100 \tag{38}$$

Recent studies [42] with VLA have found 13 massive black holes in dwarf galaxies that are less than a billion light-years from Earth. Their study indicates that roughly half of the massive black holes in dwarf galaxies are not in the centres of those galaxies. This probably indicates that the galaxies likely have merged with others earlier in their history, as elaborated in this work, thus maintaining the validity of this argument.

4. Extremal Black Holes

Irrespective of the composition of black holes (either baryonic or admixed with dark matter), the extremal rotating and charged black holes will exhibit unique properties with respect to the ratios of angular momentum to entropy, charge to entropy, etc. Irrespective of the mass of the black hole, holography implies a phase space associated with the interior volume which remarkably turns out to be just one quantum of phase space filled with just one photon: [43]; [44]

$$\left(d^3p\right)\left(d^3x\right)_{insidehorizon} = \left(\frac{\hbar c^2}{GM}\right)^3 \left(\frac{GM}{c^2}\right)^3 = \hbar^3 \tag{39}$$

For any black hole [35][44][45], we have Entropy × horizon curvature = $k_B \frac{c^3}{\hbar G}$

For extreme rotating black holes whose angular momentum (for any mass M) is given as

$$J = \frac{GM^2}{c} \tag{40}$$

The area of the horizon is:

$$A = \frac{G^2M^2}{c^4} \tag{41}$$

So that the angular momentum per unit area is:

$$\sigma = \frac{J}{=}\frac{c^3}{G} \tag{42}$$

Now the entropy per unit area, works out as:

$$S = k_B \frac{GM^2}{\hbar c} / \frac{G^2M^2}{c^4} \approx k_B \frac{c^3}{\hbar G} \tag{43}$$

(where the black hole entropy is $\sim \frac{GM^2}{\hbar c}$) Equations (42) and (43), imply the remarkable relation that for extremal black holes (of any mass), the ratio of angular momentum to entropy, i.e.,

$$\frac{J}{S} = \frac{\hbar}{k_B} \tag{44}$$

This is just the same as for an elementary particle or quantum of radiation (spin $\hbar$, entropy k_B)

Thus the black hole interior volume (for whatever black hole mass in any number of spatial dimensions) was earlier shown to be associated with a phase space of just $\hbar^3$ (having a single photon). Now the ratio of angular momentum to entropy is just $\frac{\hbar}{k_B}$ Also for extremely charged black holes:

$$Q^2 = GM^2 \tag{45}$$

This implies that the ratio of charge to angular momentum for extremal black holes (of any mass) is just:

$$\frac{Q^2}{J} = c \tag{46}$$

Or a dimensionless 'black hole fine structure' constant of:

$$\frac{Q^2}{Jc} = 1 \tag{47}$$

(analogous to $\alpha = \frac{e^2}{\hbar c} or \frac{g^2}{\hbar c} \sim 1$ in case of strong interaction)

This universal relation, $\frac{J}{Q^2} = \frac{1}{c}$ is similar to the von-Klitzing constant $\frac{\hbar}{e^2}$ is similar to the von-Klitzing constant $\frac{e^2}{\hbar}$, in superconductors (it was earlier indicated that black holes are diamagnetic, excluding flux, like superconductors). The charge to entropy ratio is:

$$\frac{Q^2}{S} = \frac{\hbar c}{k_B} \tag{48}$$

All the above relations involve only the fundamental constants, $\hbar$, c, k_B are even independent of G, although black holes are strongly gravitating systems. Several similar properties of black holes corresponding to those of elementary particles including the fact that the gyromagnetic ratio of Kerr-Newman extremal black holes is 2, just the same as a Dirac particle were indicated earlier [46]

Conclusion

We have considered the formation of supermassive black holes that seem to pervade at all epochs, and are present at the centres of large galaxies, and clusters. We have analysed the dynamics of intermediate mass black holes, and conclude that the mergers of such IMBHs is not a viable mechanism for the formation of supermassive black holes. Considering the presence of billion solar mass black holes in the early Universe (at larger redshifts), we propose a suitable model for their formation invoking dark matter. Dark matter, which is supposed to constitute about 25 per cent of the energy density of the Universe, does not interact with radiation. And as a consequence can collapse without being stopped by radiation at very early epochs. These can then provide seeds for the formation of SMBH at large redshifts.

We have also looked at the possible connection between the gravitational binding energies of large galaxies, etc., and the masses of their central black holes. Using this relation (between the gravitational binding energy of the host structure and the black hole energy), we have shown why globular clusters are unlikely to harbour large black holes and why dwarf galaxies, if they have to host black holes, should have observed mass-to-light ratios of ~ 100.

We have also discussed some interesting properties of extremal black holes of any mass. For extremal rotating and charged black holes we obtain unique values for ratios of angular momentum to entropy, charge to entropy, etc. It turns out that these ratios can be expressed in terms of fundamental constants in nature, having analogies with other physical systems, like in condensed matter physics

References

[1] C. Sivaram. *Proceedings of the 119th symposium of the international astronomical union*, page 509, 1985.

[2] C. Sivaram, K. Arun, and O. V. Kiren. Forming supermassive black holes like j1342+0928 (invoking dark matter) in early universe. *Astrophysics and Space Science*, 363(2):40, Jan 2018.

[3] C. Sivaram and K. Arun. Intermediate mass black holes: Their motion and associated energetics. *Advances in High Energy Physics*, 2014:924848, Feb 2014.

[4] P. Hut J. Makino S. F. P. Zwart, H. Baumgardt and S. L. W. McMillan. Formation of massive black holes through runaway collisions in dense young star clusters,. *Nature*, 428(6984):724726, 2004.

[5] Piero Madau and Martin J. Rees. Massive black holes as population III remnants. *The Astrophysical Journal*, 551(1):L27–L30, apr 2001.

[6] A. Heger and S. E. Woosley. The nucleosynthetic signature of population III. *The Astrophysical Journal*, 567(1):532–543, mar 2002.

[7] Tiziana Di Matteo, Jrg Colberg, Volker Springel, Lars Hernquist, and Debora Sijacki. Direct cosmological simulations of the growth of black holes and galaxies. *The Astrophysical Journal*, 676(1):33–53, mar 2008.

[8] M. J. Jee, H. C. Ford, G. D. Illingworth, R. L. White, T. J. Broadhurst, D. A. Coe, G. R. Meurer, A. van der Wel, N. Benitez, J. P. Blakeslee, R. J. Bouwens, L. D. Bradley, R. Demarco, N. L. Homeier, A. R. Martel, and S. Mei. Discovery of a ringlike dark matter structure in the core of the galaxy cluster cl 0024+17. *The Astrophysical Journal*, 661(2):728–749, jun 2007.

[9] James Binney and Scott Tremaine. *Galactic dynamics*. 1987.

[10] Eduardo Bañados, Bram P. Venemans, Chiara Mazzucchelli, Emanuele P. Farina, Fabian Walter, Feige Wang, Roberto Decarli, Daniel Stern, Xiaohui Fan, Frederick B. Davies, Joseph F. Hennawi, Robert A. Simcoe, Monica L. Turner, Hans-Walter Rix, Jinyi Yang, Daniel D. Kelson, Gwen C. Rudie, and Jan Martin Winters. An 800-million-solar-mass black hole in a significantly neutral universe at a redshift of 7.5. *Nature*, 553(7689):473–476, dec 2017.

[11] Bram P. Venemans, Fabian Walter, Roberto Decarli, Eduardo Bañados, Chris Carilli, Jan Martin Winters, Karl Schuster, Elisabete da Cunha, Xiaohui Fan, Emanuele Paolo Farina, Chiara Mazzucchelli, Hans-Walter Rix, and Axel Weiss. Copious amounts of dust and gas in a z = 7.5 quasar host galaxy. *The Astrophysical Journal*, 851(1):L8, dec 2017.

[12] C. Sivaram and K. Arun. Some aspects of intermediate mass black holes. *arXiv.0709.4597*, 2007.

[13] C Sivaram. A general upper limit on the mass and entropy production of a cluster of supermassive objects. *Astrophysics and Space Science*, 86(2):501–504, 1982.

[14] Stuart L. Shapiro and Saul A. Teukolsky. *Black holes, white dwarfs, and neutron stars : the physics of compact objects*. 1983.

[15] de Sabbata and C. Sivaram. Nato advanced study institutes series, "quantum mechanics in curved space-time". B230:485, 1990.

[16] S. L. Shapiro and S. A. Teukolsky. Black holes, white dwarfs, and neutron stars: The physics of compact objects. 1983.

[17] C. Sivaram and K. Arun. New class of dark matter objects and their detection. *Open Astronomy Journal*, 4:57, 2011.

[18] Angela S Cotera, Edwin F Erickson, Sean WJ Colgan, Janet P Simpson, David A Allen, and Michael G Burton. The discovery of hot stars near the galactic center thermal radio filaments. *The Astrophysical Journal*, 461:750, 1996.

[19] Tetsuya Nagata, Charles E Woodward, Mark Shure, and Naoto Kobayashi. Object 17: Another cluster of emission-line stars near the galactic center. *The Astronomical Journal*, 109:1676–1681, 1995.

[20] Riccardo Giacconi, Lex Kaper, EPJ van den Heuvel, et al. *Black Holes in Binaries and Galactic Nuclei: Diagnostics, Demography and Formation: Proceedings of the ESO Workshop Held at Garching, Germany, 6-8 September 1999, in Honour of Riccardo Giacconi*. Springer Science & Business Media, 2001.

[21] Fulvio Melia. *The Galactic Supermassive Black Hole*. 04 2007.

[22] Ya. B. Zeldovich and I. D. Novikov. Relativistic astrophysics, part i. 1971.

[23] John Kormendy and Douglas Richstone. Inward boundthe search for supermassive black holes in galactic nuclei. *Annual Review of Astronomy and Astrophysics*, 33(1):581–624, 1995.

[24] John Magorrian, Scott Tremaine, Douglas Richstone, Ralf Bender, Gary Bower, Alan Dressler, SM Faber, Karl Gebhardt, Richard Green, Carl Grillmair, et al. The demography of massive dark objects in galaxy centers. *The Astronomical Journal*, 115(6):2285, 1998.

[25] Kayhan Gültekin, Douglas O Richstone, Karl Gebhardt, Tod R Lauer, Scott Tremaine, Monique C Aller, Ralf Bender, Alan Dressler, SM Faber, Alexei V Filippenko, et al. The m–σ and m–l relations in galactic bulges,

and determinations of their intrinsic scatter. *The Astrophysical Journal*, 698(1):198, 2009.

[26] A Feoli and L Mancini. A hertzsprung–russell-like diagram for galaxies: The m versus $m_g\ \sigma^2$ relation. *The Astrophysical Journal*, 703(2):1502, 2009.

[27] John Kormendy. Brightness distributions in compact and normal galaxies. ii-structure parameters of the spheroidal component. *The Astrophysical Journal*, 218:333–346, 1977.

[28] Joseph Silk and Martin J Rees. Quasars and galaxy formation. *arXiv preprint astro-ph/9801013*, 1998.

[29] Tiziana Di Matteo, Volker Springel, and Lars Hernquist. Energy input from quasars regulates the growth and activity of black holes and their host galaxies. *Nature*, 433(7026):604–607, Feb 2005.

[30] Marta Volonteri and Priyamvada Natarajan. Journey to the MBH relation: the fate of low-mass black holes in the Universe. *Monthly Notices of the Royal Astronomical Society*, 400(4):1911–1918, 12 2009.

[31] C. Sivaram and K. Arun. Is the binding energy of galaxies related to their core black hole mass? *arXiv.1006.5191*, 2010.

[32] A Burrows. Neutrinos from supernova explosions. *Annual Review of Nuclear and Particle Science*, 40(1):181–212, 1990.

[33] John N Bahcall. *Neutrino astrophysics.* Cambridge University Press, 1989.

[34] C. Sivaram and K. Arun. Is the binding energy of galaxies related to their core black hole mass? *Journal of Astrophysics and Astronomy*, 42(2):12, Apr 2021.

[35] Kaushala Bandara, David Crampton, and Luc Simard. A relationship between supermassive black hole mass and the total gravitational mass of the host galaxy. *The Astrophysical Journal*, 704(2):1135, 2009.

[36] G Meylan, A Sarajedini, P Jablonka, SG Djorgovski, T Bridges, and RM Rich. Mayall ii= g1 in m31: Giant globular cluster or core of a dwarf elliptical galaxy? *The Astronomical Journal*, 122(2):830, 2001.

[37] M. Coleman Miller and Douglas P. Hamilton. Production of intermediate-mass black holes in globular clusters. *Monthly Notices of the Royal Astronomical Society*, 330(1):232–240, 02 2002.

[38] R. K. Young. Review of Publications - Principles of Stellar Dynamics, by S. Chandrasekhar. *jrasc*, 36:377, October 1942.

[39] Dheeraj R Pasham, Tod E Strohmayer, and Richard F Mushotzky. A 400-solar-mass black hole in the galaxy m82. *Nature*, 513(7516):74–76, 2014.

[40] Sandra M Faber and JS Gallagher. Masses and mass-to-light ratios of galaxies. *Annual review of astronomy and astrophysics*, 17(1):135–187, 1979.

[41] Edward W Kolb and Michael S Turner. *The early universe*. CRC press, 2018.

[42] Amy E Reines, James J Condon, Jeremy Darling, and Jenny E Greene. A new sample of (wandering) massive black holes in dwarf galaxies from high-resolution radio observations. *The Astrophysical Journal*, 888(1):36, 2020.

[43] C. Sivaram and K. Arun. Curvature, phase space, holography and black hole entropy. *International Journal of Modern Physics D*, 18(14):2167–2171, 2009.

[44] C. Sivaram and K. Arun. Some enigmatic aspects of the early universe. *Astrophysics and Space Science*, 334(2):225–230, 2011.

[45] C. Sivaram and K. Arun. Some unique constants associated with extremal black holes. *Astrophysics and Space Science*, 338(1):1–2, Mar 2012.

[46] C Sivaram and KP Sinha. Strong gravity, black holes, and hadrons. *Physical Review D*, 16(6):1975, 1977.

Chapter 2

The Quantum Black Holes and QFT Applicability Boundary

Alexander E. Shalyt-Margolin*
Institute for Nuclear Problems, Belarusian State University, Minsk, Belarus

Abstract

In the first part of this Chapter the results previously obtained by the author due to the use of the concept of **measurability** in QFT and its different applications are outlined. Then the results are applied to study the black hole thermodynamics, in particular thermodynamics of quantum black holes (qbh). In the second part of this Chapter the qbh thermodynamics is used to study the applicability boundary of the well-known QFT. It is noted that, provided the space-time foam (or same the quantum foam) at Planck's scales comprises qbh, the applicability boundary of QFT lies at the energies considerably lower than the Planck energies.

Keywords: canonical quantum field theory, quantum black holes, natural ultraviolet boundary

PACS: 11.10.-z,11.15.Ha,12.38.Bx

---

*Corresponding Author's Email: a.shalyt@mail.ru, alexm@hep.by.

In: A Guide to Black Holes
Editor: Kenath Arun
ISBN: 979-8-88697-163-7

1. Introduction

In the first part of this Section the author gives a brief outline of his previous results [1] – [6] associated with the concept of **measurability** in QFT and its different applications, specifically to thermodynamics of black holes. This concept leads to discreteness of QFT but at the energies which are considerably lower than the Planck energies $E \ll E_p$ a **measurable** QFT is nearly continuous. First, a **measurable** QFT is considered within the scope of the involved Generalized Uncertainty Principle (GUP) and minimal length on the order of Planck's $l_{min} \propto l_p$. However, in Section 4 it is shown that these approaches may be generalized to the case when there is no $l_{min} \neq 0$.

In the second part of this Section the results from the black hole thermodynamics are applied to study the Quantum Field Theory Applicability Boundary in the well-known case of continuous space-time in both the perturbative and non-perturbative regimes. It is noted that the results of D. Singleton and S. Wilburn obtained in 2011 and associated with violation of the equivalence principle in a large (classical) black hole field may be generalized to the case of quantum (Planck) black holes. It is important that these results are also valid for the **measurable** (discrete) theories.

2. Minimal Length, Minimal Inverse Temperature and Measurability

2.1. Generalized Uncertainty Principles in Quantum Theory and Thermodynamics

In this Subsection the author presents some of the results from Section 2 of the chapter [7], because they are important for this work.

It is well known that in thermodynamics an inequality for the "interior energy - inverse temperature" that is completely analogous to the standard uncertainty relation in quantum mechanics [8] can be written [11] – [16]. The only (but essential) difference of this inequality from the quantum mechanical one is that the main quadratic fluctuation is defined by means of the classical partition function rather than by the quantum mechanical expectation values. In the last years a lot of papers appeared, where the conventional momentum-coordinate uncertainty relation has been modified at very high energies on the order of the Planck energy E_p [17]–[28]. In this note we propose simple reasons for modifying the thermodynamic uncertainty relation at Planck energies. This modification results in existence of the minimal possible main quadratic fluctuation of the inverse temperature. Of course, we assume that all the thermodynamic quantities used are properly defined so that they have physical sense at such high energies. We start with the conventional Heisenberg Uncertainty Principle (relation) [8] for momentum - coordinate:

$$\Delta x \geq \frac{\hbar}{\Delta p}. \tag{1}$$

It was shown that at the Planck scale a high-energy term must appear:

$$\Delta x \geq \frac{\hbar}{\Delta p} + \alpha' l_p^2 \frac{\Delta p}{\hbar} \tag{2}$$

where l_p is the Planck length $l_p^2 = G\hbar/c^3 \simeq 1,6\ 10^{-35}m$ and α' is a constant. In [17] this term is derived from the string theory, in [20] it follows from the simple estimates of Newtonian gravity and quantum mechanics, in [24] it comes from the black hole physics, other methods can also be used [23],[25],[26]. Relation (2) is quadratic in Δp

$$\alpha' l_p^2 (\Delta p)^2 - \hbar\, \Delta x \Delta p + \hbar^2 \leq 0 \tag{3}$$

and therefore leads to the fundamental length

$$\Delta x_{min} = 2\sqrt{\alpha'} l_p. \tag{4}$$

Inequality (2) is called the Generalized Uncertainty Principle (GUP) in Quantum Theory.

Using relations (2) it is easy to obtain a similar relation for the "energy-time" pair. Indeed, (2) gives

$$\frac{\Delta x}{c} \geq \frac{\hbar}{\Delta p c} + \alpha' l_p^2 \frac{\Delta p}{c\hbar}, \tag{5}$$

then

$$\Delta t \geq \frac{\hbar}{\Delta E} + \alpha' \frac{l_p^2}{c^2} \frac{\Delta p c}{\hbar} = \frac{\hbar}{\Delta E} + \alpha' t_p^2 \frac{\Delta E}{\hbar}, \tag{6}$$

where the smallness of l_p is taken into account, so that the difference between ΔE and $\Delta(pc)$ can be neglected, and t_p is the Planck time $t_p = l_p/c = \sqrt{G\hbar/c^5} \simeq 0,54\ 10^{-43} sec.$ Inequality (6) gives, analogously to (2), the lower boundary for time $\Delta t \geq 2t_p$, determining the fundamental time

$$t_{min} = 2\sqrt{\alpha'} t_p. \tag{7}$$

Thus, the inequalities discussed can be rewritten in the standard form

$$\left\{ \begin{array}{ll} \Delta x & \geq \frac{\hbar}{\Delta p} + \alpha' \left(\frac{\Delta p}{P_{pl}}\right) \frac{\hbar}{P_{pl}} \\ \\ \Delta t & \geq \frac{\hbar}{\Delta E} + \alpha' \left(\frac{\Delta E}{E_p}\right) \frac{\hbar}{E_p}, \end{array} \right. \tag{8}$$

where $P_{pl} = E_p/c = \sqrt{\hbar c^3/G}$. Now we consider the thermodynamic uncertainty relations between the inverse temperature and interior energy of a macroscopic ensemble

$$\Delta \frac{1}{T} \geq \frac{k_B}{\Delta U}, \tag{9}$$

where k_B is the Boltzmann constant.

N. Bohr [9] and W. Heisenberg [10] first pointed out that such a kind of uncertainty principle should take place in thermodynamics. The thermodynamic uncertainty relations (9) were proved by many authors and in various ways [11] – [16]. Therefore, their validity raises no doubts. Nevertheless, relation (9) was proved in view of the standard model of the infinite-capacity heat bath encompassing the ensemble. But it is obvious from the above inequalities that at very high energies the capacity of the heat bath can no longer be assumed infinite at the Planck scale. Indeed, the total energy of the pair "heat bath - ensemble" may be arbitrary large but finite, merely as the Universe was born at a finite energy. Hence the quantity that can be interpreted as the temperature of the ensemble must have the upper limit and so does its main quadratic deviation. In other words, the quantity $\Delta(1/T)$ must be bounded from below. But in this case an additional term should be introduced into (9)

$$\Delta \frac{1}{T} \geq \frac{k_B}{\Delta U} + \eta \Delta U, \tag{10}$$

where η is a coefficient. Dimension and symmetry reasons give

$$\eta \sim \frac{k_B}{E_p^2} \quad or \quad \eta = \alpha' \frac{k_B}{E_p^2}. \tag{11}$$

As in the previous cases, inequality (10) leads to the fundamental (inverse) temperature

$$T_{max} = \frac{\hbar}{2\sqrt{\alpha'}t_p k_B} = \frac{E_p}{2\sqrt{\alpha'}k_B} = \frac{T_p}{2\sqrt{\alpha'}} = \frac{\hbar}{t_{min} k_B},$$
$$\beta_{min} = \frac{1}{k_B T_{max}} = \frac{t_{min}}{\hbar}. \tag{12}$$

In the work [29] the black hole horizon temperature has been measured with the use of the Gedanken experiment. In the process the Generalized Uncertainty Relations in Thermodynamics (10) have been derived also. Expression (10) has been considered in the monograph [30] within the scope of the mathematical physics methods.

Thus, we obtain a system of the generalized uncertainty relations in the symmetric form

$$\left\{ \begin{array}{lcl} \Delta x & \geq & \frac{\hbar}{\Delta p} + \alpha' \left(\frac{\Delta p}{P_{pl}}\right) \frac{\hbar}{P_{pl}} + \ldots \\ \Delta t & \geq & \frac{\hbar}{\Delta E} + \alpha' \left(\frac{\Delta E}{E_p}\right) \frac{\hbar}{E_p} + \ldots \\ \Delta \frac{1}{T} & \geq & \frac{k_B}{\Delta U} + \alpha' \left(\frac{\Delta U}{E_p}\right) \frac{k_B}{E_p} + \ldots \end{array} \right. \tag{13}$$

Or, in the equivalent form, we have

$$\left\{ \begin{array}{lcl} \Delta x & \geq & \frac{\hbar}{\Delta p} + \alpha' l_p^2 \frac{\Delta p}{\hbar} + \ldots \\ \Delta t & \geq & \frac{\hbar}{\Delta E} + \alpha' t_p^2 \frac{\Delta E}{\hbar} + \ldots \\ \Delta \frac{1}{T} & \geq & \frac{k_B}{\Delta U} + \alpha' \frac{1}{T_p^2} \frac{\Delta U}{k_B} + \ldots, \end{array} \right. \tag{14}$$

where the dots mean the existence of higher-order corrections as in [31]. Here T_p is the Planck temperature: $T_p = E_p/k_B$.

In literature the relation (9) is referred to as the Uncertainty Principle in Thermodynamics (UPT). Let us call the relation (10) the Generalized Uncertainty Principle in Thermodynamics (GUPT).

In this case, without the loss of generality and for symmetry, it is assumed that a dimensionless constant in the right-hand side of GUP (formula (2)) and in the right-hand side of GUPT (formula (10)) is the same – α'.

2.2. Minimal Length and Measurability Notion in Quantum Theory

First, we consider in this Subsection the principal definitions from [1],[2]which are required to derive the key formulae in the second part of the Subsection and to obtain further results.

Definition I. Let us call as **primarily measurable variation** any small variation (increment) $\widetilde{\Delta} x_{\mu}$ of any spatial coordinate x_{μ} of the arbitrary point $x_{\mu}, \mu = 1, ..., 3$ in some space-time system *R* if it may be realized in the form of the uncertainty (standard deviation) Δx_{μ}, when this coordinate is measured within the scope of Heisenberg's Uncertainty Principle (HUP) [8] (formula (1) in the general case):

$$\widetilde{\Delta} x_{\mu} = \Delta x_{\mu}, \Delta x_{\mu} \simeq \frac{\hbar}{\Delta p_{\mu}}, \mu = 1, 2, 3 \tag{15}$$

for some $\Delta p_{\mu} \neq 0$.

Similarly, at $\mu = 0$ for the pair "time-energy" (t, E), let us call any small variation (increment) the **primarily measurable variation** in the value of time $\widetilde{\Delta} x_0 = \widetilde{\Delta} t_0$ if it may be realized in the form of the uncertainty (standard deviation) $\Delta x_0 = \Delta t$ and then

$$\widetilde{\Delta} t = \Delta t, \Delta t \simeq \frac{\hbar}{\Delta E} \tag{16}$$

for some $\Delta E \neq 0$. Formula (16) is nothing else but formula (6) for $\Delta E \ll E_p$.

Here HUP is given for the non-relativistic case. In the relativistic case HUP has the distinctive features [32] which, however, are of no significance for the general formulation of **Definition I.**,being associated only with particular alterations in the right-hand side of the second relation (16).

It is clear that at low energies $E \ll E_P$ (momenta $P \ll P_{pl}$) **Definition I.** sets a lower bound for the **primarily measurable variation** $\widetilde{\Delta} x_\mu$ of any space-time coordinate x_μ.
At high energies E (momenta P) this is not the case if E (P) has no upper limit. But, according to the modern knowledge, E (P) is bounded by some maximal quantities E_{max}, (P_{max})

$$E \leq E_{max}, P \leq P_{max}, \tag{17}$$

where, in general, E_{max}, P_{max} may be on the order of the Planck quantities $E_{max} \propto E_P$, $P_{max} \propto P_{pl}$ and also may be the trans-Planck's quantities.

In any case the quantities P_{max} and E_{max} lead to the introduction of the minimal length l_{min} and of the minimal time t_{min}.

Supposition II. There is the minimal length l_{min} as *a minimal measurement unit* for all **primarily measurable variations** having the dimension of length, whereas the minimal time $t_{min} = l_{min}/c$ as *a minimal measurement unit* for all quantities or **primarily measurable variations (increments)** having the dimension of time, where c is the speed of light.

l_{min} and t_{min} are naturally introduced as $\Delta x_\mu, \mu = 1, 2, 3$ and Δt in Equations (15) and (16) for $\Delta p_\mu = P_{max}$ and $\Delta E = E_{max}$.

For definiteness, we consider that E_{max} and P_{max} are the quantities on the order of the Planck quantities, then l_{min} and t_{min} are also on the order of the Planck quantities $l_{min} \propto l_P$, $t_{min} \propto t_P$.

Definition I. and **Supposition II.** are quite natural in the sense that there are no physical principles with which they are inconsistent.
The combination of **Definition I.** and **Supposition II.** will be called the **Principle of Bounded Primarily Measurable Space-Time Variations (Increments)** or, for short, the **Principle of Bounded Space-Time Variations (Increments)** with abbreviation (PBSTV).
As the minimal unit of measurement l_{min} is available for all the **primarily measurable variations** ΔL having the dimensions of length, the "Integrality Condition" (IC) is the case

$$\Delta L = N_{\Delta L} l_{min}, \tag{18}$$

where $N_{\Delta L} > 0$ is an integer number.

In a like manner, the same “Integrality Condition” (IC) is the case for all the **primarily measurable variations** Δt having the dimensions of time. And similar to Equation (18), we get the following expression for any time Δt:

$$\Delta t \equiv \Delta t(N_t) = N_{\Delta t} t_{min}, \tag{19}$$

where similarly $N_{\Delta t} > 0$ is an integer number too.

Definition 1 (Primary or Elementary Measurability).
(1) *In accordance with the PBSTV, let us define the quantity having the dimension of length or time as* **primarily (or elementarily) measurable**, *when it satisfies the Equation (18) (and respectively Equation (19)).*
(2) *Let us define any physical quantity as* **primarily (or elementarily) measurable**, *when its value is consistent with point (1) of this Definition.*

It is convenient to use the deformation parameter α_a. This parameter has been introduced earlier in the papers [33],[7],[34]–[37] as a *deformation parameter* (according to paper [38]), on going from the canonical quantum mechanics to the quantum mechanics at Planck's scales (Early Universe) that is considered to be the quantum mechanics with a minimal length (QMML):

$$\alpha_a = l_{min}^2/a^2, \tag{20}$$

where a is the measuring scale. It is easily seen that the parameter α_a from Equation (20) is discrete as it is nothing else but

$$\alpha_a = l_{min}^2/a^2 = \frac{l_{min}^2}{N_a^2 l_{min}^2} = \frac{1}{N_a^2}. \tag{21}$$

At the same time, from Equation (21) it is evident that α_a is irregularly discrete.

It should be noted that physical quantities complying with **Definition 1** are inadequate for the research of physical systems.

Indeed, such a variable as

$$\alpha_{N_a l_{min}}(N_a l_{min}) = p(N_a)\frac{l_{min}^2}{\hbar} = l_{min}/N_a, \tag{22}$$

where $\alpha_{N_a l_{min}} = \alpha_a$ is taken from formula (21 at $a = N_a l_{min}$ and $p(N_a) = \frac{\hbar}{N_a l_{min}}$ is the corresponding **primarily measurable** momentum), is fully expressed in terms of *only* **Primarily Measurable**

Quantities of **Definition 1** and hence it can appear at any stage of calculations but, apparently, is not complying with **Definition 1**. Because of this, it is necessary to introduce the following definition generalizing **Definition 1**:

Definition 2. Generalized Measurability.
We can call any physical quantity as **generalized-measurable** or, for simplicity, **measurable** if any of its values may be obtained in terms of **Primarily Measurable Quantities** of **Definition 1**.

In what follows, for simplicity, we use the term **Measurability** instead of **Generalized Measurability**.

It is evident that any **primarily measurable quantity (PMQ)** is **measurable**. Generally speaking, the contrary is not correct as indicated by formula (22).

The **generalized-measurable** quantities follow from the **Generalized Uncertainty Principle (GUP)** (formula (2)), naturally leading to the minimal length l_{min} [17]–[28]:

$$\Delta x_{min} = 2\sqrt{\alpha'} l_p \doteq l_{min}. \tag{23}$$

For convenience, we denote the minimal length $l_{min} \neq 0$ by ℓ and $t_{min} \neq 0$ by $\tau = \ell/c$. Solving inequality (2), in the case of equality we obtain the apparent formula

$$\Delta p_{\pm} = \frac{(\Delta x \pm \sqrt{(\Delta x)^2 - 4\alpha' l_p^2})\hbar}{2\alpha' l_p^2}. \tag{24}$$

Next, into this formula we substitute the right-hand part of formula (18) for $L = x$. Considering (23), we can derive the following:

$$\begin{aligned} \Delta p_{\pm} = \frac{(N_{\Delta x} \pm \sqrt{(N_{\Delta x})^2 - 1})\hbar\ell}{\frac{1}{2}\ell^2} = \\ = \frac{2(N_{\Delta x} \pm \sqrt{(N_{\Delta x})^2 - 1})\hbar}{\ell}. \end{aligned} \tag{25}$$

But it is evident that at low energies $E \ll E_p; N_{\Delta x} \gg 1$ the plus sign in the nominator (25) leads to the contradiction as this results in very high (much greater than the Planck) values of Δp. Because of this, it is necessary to select the minus sign in the numerator (25). Then, multiplying the left and right sides of (25) by the same number $N_{\Delta x} + \sqrt{N_{\Delta x}^2 - 1}$, we get

$$\Delta p = \frac{2\hbar}{(N_{\Delta x} + \sqrt{N_{\Delta x}^2 - 1})\ell}. \tag{26}$$

Δp from formula (26) is the **generalized-measurable** quantity in the sense of **Definition 2.** However, it is clear that at low energies $E \ll E_p$, i.e. for $N_{\Delta x} \gg 1$, we have $\sqrt{N_{\Delta x}^2 - 1} \approx N_{\Delta x}$. Moreover, we have

$$\lim_{N_{\Delta x} \to \infty} \sqrt{N_{\Delta x}^2 - 1} = N_{\Delta x}. \tag{27}$$

Therefore, in this case (26) may be written as follows:

$$\Delta p \doteq \Delta p(N_{\Delta x}, HUP)$$
$$= \frac{\hbar}{1/2(N_{\Delta x} + \sqrt{N_{\Delta x}^2 - 1})\ell} \approx \frac{\hbar}{N_{\Delta x}\ell} = \frac{\hbar}{\Delta x}; N_{\Delta x} \gg 1, \tag{28}$$

in complete conformity with HUP. Besides, $\Delta p \doteq \Delta p(N_{\Delta x}, HUP)$, to a high accuracy, is a **primarily measurable** quantity in the sense of **Definition 1**.

And, vice versa, it is obvious that at high energies $E \approx E_p$, i.e., for $N_{\Delta x} \approx 1$, there is no way to transform formula (26). So we can write

$$\Delta p \doteq \Delta p(N_{\Delta x}, GUP) = \frac{\hbar}{1/2(N_{\Delta x} + \sqrt{N_{\Delta x}^2 - 1})\ell}; N_{\Delta x} \approx 1. \tag{29}$$

At the same time, $\Delta p \doteq \Delta p(N_{\Delta x}, GUP)$ is a **Generalized Measurable** quantity in the sense of **Definition 2**.

Thus, we have

$$GUP \to HUP \tag{30}$$

for

$$(N_{\Delta x} \approx 1) \to (N_{\Delta x} \gg 1). \tag{31}$$

Also, we have

$$\Delta p(N_{\Delta x}, GUP) \to \Delta p(N_{\Delta x}, HUP), \tag{32}$$

where $\Delta p(N_{\Delta x}, GUP)$ is taken from formula (29), whereas $\Delta p(N_{\Delta x}, HUP)$ – from formula (28).

Comment 2.*
From the above formulae it follows that, within GUP, the **primarily measurable** *variations (quantities) are derived to a high accuracy from the* **generalized-measurable** *variations (quantities) only in the low-energy limit* $E \ll E_P$.

Next, within the scope of GUP, we can correct a value of the parameter α_a from formula (21) substituting a for Δx in the expression $1/2(N_{\Delta x} + \sqrt{N_{\Delta x}^2 - 1})\ell$.

Then, at low energies $E \ll E_p$, we have the **primarily measurable** quantity $\alpha_a(HUP)$

$$\alpha_a \doteq \alpha_a(HUP) = \frac{1}{[1/2(N_a + \sqrt{N_a^2 - 1})]^2} \approx \frac{1}{N_a^2}; N_a \gg 1, \tag{33}$$

that, to a high accuracy, corresponds to the value from formula (21).

Accordingly, at high energies, we have $E \approx E_p$

$$\alpha_a \doteq \alpha_a(GUP) = \frac{1}{[1/2(N_a + \sqrt{N_a^2 - 1})]^2}; N_a \approx 1. \tag{34}$$

When going from high energies $E \approx E_p$ to low energies $E \ll E_p$, we can write

$$\alpha_a(GUP) \stackrel{(N_a \approx 1) \rightarrow (N_a \gg 1)}{\longrightarrow} \alpha_a(HUP) \tag{35}$$

in complete conformity to *Comment 2**.

2.3. Minimal Inverse Temperature and Measurability

Now, let us return to the thermodynamic relation (10) in the case of equality:

$$\Delta \frac{1}{T} = \frac{k_B}{\Delta U} + \eta \, \Delta U, \tag{36}$$

that is equivalent to the quadratic equation

$$\eta \, (\Delta U)^2 - \Delta \frac{1}{T} \Delta U + k_B = 0. \tag{37}$$

The discriminant of this equation, with due regard for formula (11), is equal to

$$D = (\Delta\frac{1}{T})^2 - 4\eta k_B = (\Delta\frac{1}{T})^2 - 4\alpha'\frac{k_B^2}{E_p^2} \geq 0, \qquad (38)$$

leading directly to $(\Delta\frac{1}{T})_{min}$

$$(\Delta\frac{1}{T})_{min} = 2\sqrt{\alpha'}\frac{k_B}{E_p} \qquad (39)$$

or, due to the fact that k_B is constant, we have

$$(\Delta\frac{1}{k_B T})_{min} = \frac{2\sqrt{\alpha'}}{E_p}. \qquad (40)$$

It is clear that $(\Delta\frac{1}{T})_{min}$ corresponds to T_{max} from formula (12)

$$T_{max} \approx T_p \gg 0. \qquad (41)$$

In this case $\Delta\frac{1}{T} \approx \frac{1}{T}$ and, of course, we can assume that

$$(\frac{1}{T})_{min} \doteq \widetilde{\tau} = \frac{1}{T_{max}}. \qquad (42)$$

Trying to find from formula (42) a minimal unit of measurability for the inverse temperature and introducing the "Integrality Condition" (IC) in line with the conditions (18),(19)

$$\frac{1}{T} = N_{1/T}\widetilde{\tau}, \qquad (43)$$

where $N_{1/T} > 0$ is an integer number, we can introduce an analog of the **primary measurability** notion into thermodynamics.

Definition 3 (Primary Thermodynamic Measurability).
(1) *Let us define a quantity having the dimensions of inverse temperature as* **primarily measurable** *when it satisfies the relation (43).*
(2) Let us define any physical quantity in thermodynamics as **primarily measurable** *when its value is consistent with point (1) of this Definition.*

Definition 3 in thermodynamics is analogous to the **Primary Measurability** in a quantum theory (**Definition 1**).

Now we consider the quadratic equation (37) in terms of **measurable quantities** in the sense of **Definition 3**. In accordance with this definition and with formula (43) $\Delta(1/T)$, we can write

$$\Delta\frac{1}{T} = N_{\Delta(1/T)}\widetilde{\tau}, \tag{44}$$

where $N_{\Delta(1/T)} > 0$ is an integer number.

The quadratic equation (37) takes the following form:

$$\eta\,(\Delta U)^2 - N_{\Delta(1/T)}\widetilde{\tau}\Delta U + k_B = 0. \tag{45}$$

Then, due to formula (40), we can find the "**measurable**" roots of equation (45) for ΔU as follows:

$$\begin{aligned}(\Delta U)_{meas,\pm} &= \frac{[N_{\Delta(1/T)} \pm \sqrt{N^2_{\Delta(1/T)} - 1}]\widetilde{\tau}}{2\eta} = \\ &= \frac{2k_B[N_{\Delta(1/T)} \pm \sqrt{N^2_{\Delta(1/T)} - 1}]\widetilde{\tau}}{\widetilde{\tau}^2} = \\ &\quad \frac{2k_B[N_{\Delta(1/T)} \pm \sqrt{N^2_{\Delta(1/T)} - 1}]}{\widetilde{\tau}}.\end{aligned} \tag{46}$$

The last line in (46) is associated with the obvious relation $2\eta = \frac{\widetilde{\tau}^2}{2k_B}$.

In this way we derive a complete analog of the corresponding relation (25) from a quantum theory by replacement

$$\Delta p_{\pm} \Rightarrow \Delta U_{meas,\pm};\, N_{\Delta x} \Rightarrow N_{\Delta(1/T)};\, \hbar \Rightarrow k_B. \tag{47}$$

As, for **low temperatures and energies**, $T \ll T_{max} \propto T_p$, we have $1/T \gg 1/T_p$ and hence $\Delta(1/T) \gg 1/T_p$ and $N_{\Delta(1/T)} \gg 1$.

Next, in analogy with Subsection 2.2, in formula (46) we can have only the minus-sign root, otherwise, at sufficiently high $N_{\Delta(1/T)} \gg 1$ for $(\Delta U)_{meas,+}$ we can get $(\Delta U)_{meas,+} \gg E_p$. But this is impossible for low temperatures (energies).

On the contrary, the minus sign in (46) is consistent with high and low energies.

So, taking the root value in (46) corresponding to this sign and multiplying the nominator and denominator in (46) by $N_{\Delta(1/T)} + \sqrt{N^2_{\Delta(1/T)} - 1}$, we obtain

$$(\Delta U)_{meas} = \frac{2k_B}{(N_{\Delta(1/T)} + \sqrt{N^2_{\Delta(1/T)} - 1})\widetilde{\tau}} \tag{48}$$

to have a complete analog of the corresponding relation from (26) in a quantum theory by substitution according to formula (47).

Then it is clear that, in analogy with Subsection 2.2, for low energies and temperatures $N_{\Delta(1/T)} \gg 1$ (48) may be rewritten as

$$\begin{aligned}(\Delta U)_{meas} &\doteq (\Delta U)_{meas}(T \ll T_{max}) \\ &= \frac{2k_B}{(N_{\Delta(1/T)} + \sqrt{N^2_{\Delta(1/T)} - 1})\widetilde{\tau}} \approx \\ &\approx \frac{k_B}{N_{\Delta(1/T)}\widetilde{\tau}}, N_{\Delta(1/T)} \gg 1,\end{aligned} \tag{49}$$

i.e., the Uncertainty Principle in Thermodynamics (UPT, formula (9)) is involved. In this case, due to the last formula, ΔU_{meas} represents a **primarily measurable** thermodynamic quantity in the sense of **Definition 3** to a high accuracy.

Of course, at high energies the last term in the formula (49) is lacking and, for $T \approx T_{max}; N_{\Delta(1/T)} \approx 1$, we have:

$$\begin{aligned}(\Delta U)_{meas} &\doteq (\Delta U)_{meas}(T \approx T_{max}) \\ &= \frac{k_B}{1/2(N_{\Delta(1/T)} + \sqrt{N^2_{\Delta(1/T)} - 1})\widetilde{\tau}}, \\ &N_{\Delta(1/T)} \approx 1.\end{aligned} \tag{50}$$

From (50) it follows that at high temperatures (energies) $(\Delta U)_{meas}$ could hardly be a **primarily measurable** thermodynamic quantity. Because of this, it is expedient to use a counterpart of **Definition 2**.

Definition 4. Generalized Measurability in Thermodynamics. Any physical quantity in thermodynamics may be referred to as **generalized-measurable** or, for simplicity, **measurable** if any of its values may be obtained in terms of the **Primary Thermodynamic Measurability** of **Definition 3**.

In this way $(\Delta U)_{meas}$ from the formula (50) is a **measurable** quantity.

Based on the preceding formulae, it is clear that we have the limiting transition

$$(\Delta U)_{meas}(T \approx T_{max}) \stackrel{(N_{\Delta(1/T)} \approx 1) \rightarrow (N_{\Delta(1/T)} \gg 1)}{\longrightarrow} (\Delta U)_{meas}(T \ll T_{max} \propto T_p), \quad (51)$$

that is analogous to the corresponding formula (35) in a quantum theory.

Therefore, in this case the analog of *Comment 2**. in Subsection 2.2 is valid.

Comment 2 Thermodynamics*

From the above formulae it follows that, within GUPT (10), the **primarily measurable** *variations (quantities) are derived, to a high accuracy, from the* **generalized-measurable** *variations (quantities) only in the low-temperature limit* $T \ll T_{max} \propto T_p$.

To conclude this Section, it seems logical to make several important **remarks**.

R2.1. It is obvious that all the calculations associated with **measurability** of inverse temperature $\frac{1}{T}$ are valid for $\beta = \frac{1}{k_B T}$ as well. Specifically, introducing $\beta_{min} \doteq \widetilde{\beta} = \widetilde{\tau}/k_B$, we can rewrite all the corresponding formulae in the "**measurable**" variant with appropriate replacement.

R2.2. Naturally, the problem of compatibility between the **measurability** definitions in quantum theory and in thermodynamics arises: is there any contradiction between **Definition 1** from Subsection 2.2 and **Definition 3** from Subsection 2.3?

On the basis of the formulae (12) from Subsection 2.1 and (42) from Subsection 2.3 we can state:

measurability *in quantum theory and* ***thermodynamic measurability*** *are completely compatible and consistent as the minimal unit of inverse temperature* $\widetilde{\tau}$ *is nothing else but the minimal time* $t_{min} = \tau$ *up to a constant factor. And hence* $N_{1/T}$, $(N_{\Delta(1/T)})$ *is nothing else but* N_t, $(N_{\Delta t})$

in (19). Then it is clear that $N_t = N_{a=tc}$.

R2.3. Finally, from the above formulae (49), (50) it follows that the **measurable** temperature T is varying as follows:

$$T = \frac{T_{max}}{N_{1/T}}, T \ll T_{max} \propto T_p, N_{1/T} \gg 1;$$

$$T = \frac{T_{max}}{1/2(N_{1/T} + \sqrt{N_{1/T}^2 - 1})}, T \approx T_{max} \propto T_p, N_{1/T} \approx 1. \qquad (52)$$

In such a way **measurable** temperature is a **discrete quantity** but at low energies it is almost constantly varying, so the theoretical calculations are very similar to those of the well-known continuous theory. In the reality, discreteness manifests itself in the case of high energies only.

3. Black Holes and Measurability

Now let us show the applicability of the results from Section 2 to a quantum theory of black holes. Consider the case of Schwarzschild black hole. It seems logical to support the idea, suggested in the Introduction to the recent overview presented by seven authors [39]: "Since for (asymptotically flat Schwarzschild) black holes the temperatures increase as their masses decrease, soon after Hawking's discovery, it became clear that a complete description of the evaporation process would ultimately require a consistent quantum theory of gravity. This is necessary as the semi-classical formulation of the emission process breaks down during the final stages of the evaporation as characterized by Planckian values of the temperature and spacetime curvature". Naturally, it is important to study the transition from low to high energies in the indicated case.

This Section presents gravitational dynamics at low $E \ll E_p$ and at high $E \approx E_p$ energies in the case of the Schwarzschild black hole and in a more general case of the space with static spherically-symmetric horizon in space-time when using **measurable quantities** from the previous Section.

It should be noted that such spaces and even considerably more general cases have been thoroughly studied from the viewpoint of gravitational thermodynamics in remarkable works of professor T.Padmanbhan [40]–[51] (the list of references may be much longer).

First, the author has studied the above-mentioned case in [52] and from the suggested viewpoint – in [1]. But, proceeding from Section 2 of the present paper, it is possible to extend the results from [1].

In what follows we use the symbols from [51] which have been also used in [1]. The case of a static spherically-symmetric horizon in space-time is considered, the horizon being described by the metric

$$ds^2 = -f(r)c^2dt^2 + f^{-1}(r)dr^2 + r^2 d\Omega^2. \qquad (53)$$

The horizon location will be given by a simple zero of the function $f(r)$, at the radius $r = a$.

Then at the horizon $r = a$ Einstein's field equations ([51], eq.(117)) take the form

$$\frac{c^4}{G}\left[\frac{1}{2}f'(a)a - \frac{1}{2}\right] = 4\pi P a^2, \qquad (54)$$

where $P = T_r^r$ is the trace of the momentum-energy tensor and radial pressure. Therewith, the conditions $f(a) = 0$ and $f'(a) \neq 0$ must be fulfilled.

On the other hand, it is known that for horizon spaces one can introduce the temperature that can be identified with an analytic continuation to imaginary time. In the case under consideration we have ([51], eq.(116))

$$k_B T = \frac{\hbar c f'(a)}{4\pi}. \qquad (55)$$

In [51] it is shown that in the initial (continuous) theory the Einstein Equation for horizon spaces in the differential form may be written as a thermodynamic identity (the first principle of thermodynamics) ([51], formula (119)):

$$\underbrace{\frac{\hbar c f'(a)}{4\pi}}_{k_B T} \underbrace{\frac{c^3}{G\hbar} d\left(\frac{1}{4}4\pi a^2\right)}_{dS} \underbrace{- \frac{1}{2}\frac{c^4 da}{G}}_{-dE} = \underbrace{P d\left(\frac{4\pi}{3}a^3\right)}_{P\,dV}, \qquad (56)$$

where, as noted above, T – temperature of the horizon surface, S – corresponding entropy, E– internal energy, V – space volume.

It is impossible to use (56) in the formalism under consideration because, as follows from the results given in the previous section and in [1], da, dS, dE, dV are not **measurable quantities**.

First, we assume that a value of the radius r at the point a is a **primarily measurable quantity** in the sense of **Definition 1** from Subsection 2.2., i.e., $a = a_{meas} = N_a \ell$, where $N_a > 0$ - integer, and the temperature T from the left-hand side of (55) is the **measurable** temperature $T = T_{meas}$ in the sense of **Definition 3** from Subsection 2.2.3.

Then, in terms of **measurable** quantities, first we can rewrite (54) as

$$\frac{c^4}{G}\left[\frac{2\pi k_B T}{\hbar c} a_{meas} - \frac{1}{2}\right] = 4\pi P a_{meas}^2. \tag{57}$$

We express $a = a_{meas}$ in terms of the deformation parameter α_a (formula (20)) as

$$a = \ell \alpha_a^{-1/2}; \tag{58}$$

the temperature T is expressed in terms of $T_{max} \propto T_p$ from (52).

Then, considering that $T_p = E_p/k_B$, equation (57) may be given as

$$\frac{c^4}{G}[\frac{\pi E_p}{\sqrt{\alpha'} N_{1/T} \hbar c} \ell \alpha_a^{1/2} - \frac{1}{2}\alpha_a] = 4\pi P \ell^2. \tag{59}$$

Because $\ell = 2\sqrt{\alpha'} l_p$ and $l_p = \frac{\hbar c}{E_p}$, we have

$$\frac{c^4}{G}[\frac{2\pi E_p}{N_{1/T}\hbar c} l_p \alpha_a^{1/2} - \frac{1}{2}\alpha_a] = \frac{c^4}{G}[\frac{2\pi}{N_{1/T}} \alpha_a^{1/2} - \frac{1}{2}\alpha_a] = 4\pi P \ell^2. \tag{60}$$

Note that in its initial form [51] the equation (54) has been considered in a continuous theory, i.e., at low energies $E \ll E_p$. Consequently, in the present formalism it is implicitly meant that the “measurable counterpart” of the equation (54) – (57) (or the same (59),(60)) is also initially considered at low energies, in particular, $N_a \gg 1, N_{1/T} \gg 1$. Let us consider the possibility of generalizing (59),(60) to high energies taking two different cases.

3.1. *Measurable case for low energies:* $E \ll E_p$. Due to formula (28), $a = a_{meas} = N_a \ell$, where the integer number is $N_a \gg 1$ or similarly $N_{1/T} \gg 1$. In this case GUP, to a high accuracy, is extended to HUP (formula (30),(31)).

As this takes place, $\alpha_a = \alpha_a(HUP)$ is a **primarily measurable** quantity (**Definition 1**), $\alpha_a \approx N_a^{-2}$, though taking a discrete series of values but varying smoothly, in fact *continuously*. (59) is a quadratic equation, with respect to $\alpha_a^{1/2} \approx N_a^{-1}$, having the two parameters $N_{1/T}^{-1}$ and P. In this case the equation (60) may be rewritten as

$$\frac{c^4}{G}[\frac{2\pi}{N_{1/T}}\alpha_a^{1/2}(HUP) - \frac{1}{2}\alpha_a(HUP)] = 4\pi P\ell^2. \tag{61}$$

So, at low energies the equation (60) (or (61)) written for the discretely-varying α_a may be considered in a continuous theory.
As a result, in the case under study we can use the basic formulae from a continuous theory considering them valid to a high accuracy.

In particular, in the notation used for *Schwarzschild's black hole* [53], we have

$$r_s = N_a\ell = \frac{2GM}{c^2}; M = \frac{N_a\ell c^2}{2G}. \tag{62}$$

As its temperature is given by the formula

$$T_H = \frac{\hbar c^3}{8\pi GMk_B}, \tag{63}$$

at once we get

$$T_H = \frac{\hbar c}{4\pi k_B N_a \ell} = \frac{\hbar c \alpha_a^{1/2}}{4\pi k_B \ell}. \tag{64}$$

Comparing this expression to the expression with high $N_{1/T}$ ($N_{1/T} \gg 1$) for temperature from the equation (52) that is involved in (57), we can find that at low energies $E \ll E_p$, due to comment **R2.2.** from Subsection 2.3, the number $N_{1/T}$ is actually coincident with the number N_a:

$$N_{1/T} = N_a = \alpha_a^{-1/2}(HUP). \tag{65}$$

The substitution of the last expression from formula (64) into the quadratic equation (59) for $\alpha_a^{1/2}$ makes it a linear equation for α_a with a single parameter P.

3.2. *Measurable case for high energies:* $E \approx E_p$. Then, due to (29), a is the **generalized measurable** quantity $a = a_{meas} = 1/2(N_a + \sqrt{N_a^2 - 1})\ell$, with the integer $N_a \approx 1$.

The quantity

$$\Delta a_{meas}(q) = 1/2(N_a + \sqrt{N_a^2 - 1})\ell - N_a\ell = 1/2(\sqrt{N_a^2 - 1} - N_a)\ell \quad (66)$$

may be considered as a **quantum correction** for the **measurable** radius $r = a_{meas}$, that is infinitesimal at low energies $E \ll E_p$ and not infinitesimal for high energies $E \approx E_p$.

In this case there is no possibility to replace GUP by HUP. In equation (59) $\alpha_a = \alpha_a(GUP)$ is a **generalized measurable** quantity (**Definition 2**).

As noted in formula (52) of Comment **R2.3**, in this case the number $N_{1/T}$ in equation (60) is replaced by $1/2(N_{1/T} + \sqrt{N_{1/T}^2 - 1})$, i.e., the equation is of the form

$$\frac{c^4}{G}[\frac{2\pi}{1/2(N_{1/T} + \sqrt{N_{1/T}^2 - 1})}\alpha_a^{1/2}(GUP) - \frac{1}{2}\alpha_a(GUP)] = 4\pi P\ell^2. \quad (67)$$

In so doing the theory becomes really discrete, and the solutions of (67) take a discrete series of values for every N_a or $(\alpha_a(GUP))$ sufficiently close to 1.

In this formalism for a "quantum" Schwarzschild black hole (i.e., at high energies $E \approx E_p$) formula (64) is replaced by

$$T_H(Q) = \frac{\hbar c}{4\pi k_B 1/2(N_a + \sqrt{N_a^2 - 1})\ell} = \frac{\hbar c \alpha_a^{1/2}(GUP)}{4\pi k_B \ell}. \quad (68)$$

We should make several remarks which are important.

Remark 3.3. As noted in [1], the parameter $\alpha_a = \alpha_a(HUP)$, within constant factors, is coincident with the Gaussian curvature K_a [54] corresponding to the **primarily measurable** $a = N_a\ell$:

$$\alpha_a = \frac{\ell^2}{a^2} = \ell^2 K_a. \quad (69)$$

Because of this, the transition from $\alpha_a(HUP)$ to $\alpha_a(GUP)$ may be considered as a basis for "quantum corrections" to the Gaussian curvature K_a

in the high-energy region $E \approx E_p$:

$$\alpha_a(GUP) - \alpha_a(HUP) = \ell^2[\frac{1}{1/4(N_a + \sqrt{N_a^2 - 1})^2\ell^2} - \frac{1}{N_a^2\ell^2}] = \\ = \ell^2(K_a^Q - K_a), \quad (70)$$

where the "measurable quantum Gaussian curvature " K_a^Q is defined as

$$K_a^Q \doteq \frac{1}{1/4(N_a + \sqrt{N_a^2 - 1})^2\ell^2}. \quad (71)$$

In a similar way, with the use of formulae (64) and (63), we can derive a "measurable quantum correction " for the mass M of a Schwarzschild black hole at high energies.

Remark 3.4. It is readily seen that a minimal value of $N_a = 1$ is *unattainable* because in formula (29) we can obtain a value of the length l which is below the minimum $l < \ell$ for the momenta and energies above the maximal ones, and that is impossible. Thus, we always have $N_a \geq 2$. This fact was indicated in [33],[7], however, based on the other approach.

Remark 3.5. It is clear that we have the following transition:

$$Eq.(67)(E \approx E_p) \stackrel{(N_a \approx 1) \to (N_a \gg 1)}{\longrightarrow} Eq.(61)(E \ll E_p).$$

Remark 3.6. So, all the members of the gravitational equation (60) (and (67), respectively), apart from P, are expressed in terms of the measurable parameter α_a. From this it follows that P should be also expressed in terms of the measurable parameter α_a, i.e., $P = P(\alpha_a)$: $E \ll E_p$, $P = P[\alpha_a(HUP)]$ at low energies and $E \approx E_p, P = P[\alpha_a(GUP)]$ at high energies. Then, due to the above formulae, we can have for a "quantum" Schwarzschild black hole the "horizon" gravitational equation in terms of **measurable** quantities

$$(4\pi - 1)\frac{c^4}{G}\alpha_a(GUP) = 8\pi P[\alpha_a(GUP)]\ell^2, \quad (72)$$

where $\alpha_a(GUP)$ takes a discrete series of the values $\alpha_a(GUP) = (1/2(N_a + \sqrt{N_a^2 - 1}))^{-2}$; $N_a \geq 2$ is a small integer.

4. Maximal Momentum without Minimal Length

It is clear that GUP represented by formulae of (23) is a partial case of the HUP modifications (deformations) at high energies. The paper [55] reviews different approaches to the generalization of HUP, i.e., different forms of GUP with involvement of the minimal length ℓ, with and without the maximal momentum p_{max}.

However, there is one more barrier for the use of l_{min} in the theory as indicated in [27] and other works (for example, [55]). In the above-mentioned papers, it has been noted that there is a nonzero minimal uncertainty in position, i.e., l_{min} implies that there is no physical state which is the position eigenstate, since an eigenstate would, of course, have zero uncertainty in position. So, in this case a serious problem of a quantum theory is the absence of the coordinate representation in Hilbert space.

Note that in all the works l_{min} is actually (but not explicitly) introduced on the basis of some measuring procedure (different forms of the Generalized Uncertainty Principle (GUP)). In any form GUP in turn is a high-energy generalization of HUP. But in the original proof of HUP a planar geometry of the initial space-time was actively used [8]. Extension of this principle to other pairs of conjugate variables is also valid only for quantum mechanics in the planar geometry space [56]. As HUP is a local principle, at low energies in the curved space-time, by virtue of Einstein's Equivalence Principle, we can consider that in a fairly small neighborhood of any point the geometry is planar an hence HUP is valid too. But, as rule, all the results obtained point to the fact that l_{min} should be at a level of l_p, i.e., $l_{min} \propto l_p$, or even should be smaller. However, Einstein's Equivalence Principle is obviously inapplicable at the Planck scales, and there is no way to use the measuring procedure ignoring the space geometry at these scales. Meantime, none of the GUP forms [55] makes an effort to include it and is hardly completely correct.

It should be noted that many researchers are of the opinion that at very high energies (Plank's or trans-Planck's) the ultraviolet cutoff exists that is determined by some maximal momentum.

In particular, in papers [57]–[59] the author extends the heuristic approach, where it is supposed that in a theory there is a maximal bound for the measurement momenta $p = p_{max}$

$$p_{max} \doteq p_\ell = \hbar/\ell. \tag{73}$$

Here ℓ is some small length and $\tau = \ell/c$ is the corresponding time, which in [57]–[59] are respectively called the *primary* length and the *primary* time. Initially, in [2] it has been assumed that ℓ is a minimal length but in [57]–[59] it has been shown that the condition is redundant, at least when $p_{max} \propto p_{pl}$.

Without loss of generality, we can consider ℓ and τ at Plank's level, i.e., $\ell \propto l_p, \tau = \kappa t_p$, where the numerical constant κ is on the order of 1. Consequently, we have $E_\ell \propto E_p$ with the corresponding proportionality factor, where $E_\ell \doteq p_\ell c$.

In [57]–[59], in terms of ℓ, the discrete quantum deformation of canonical QFT [60]–[63] (called the **measurable** variant of QFT) is constructed, which, to a high accuracy, is coincident with the case at low energies $E \ll E_p$. A role of infinitesimal quantities $dx_i, dt, i = 1..., 3$ in such deformation is given to the quantities $\ell/N_i, \tau/N_t$, where N_i, N_t are integer numbers with the property $|N_i| \gg 1, |N_t| \gg 1$.

Passage from the **measurable** variant of QFT at low energies $E \ll E_p$ to the continuum limit, i.e., to the canonical quantum field theory, is possible in two ways: a) $\ell \to 0, \tau \to 0$ or b)$N_i \to \infty, N_t \to \infty$. As, from the start, it has been assumed that $p_{max} < \infty$, i.e., $\ell \neq 0$, in [57]–[59] initially the author has chosen variant b). The above passage and the **measurable** variant of QFT itself may be considered within the concept of the physical theory deformation [38].

It is important that all the experimentally approachable energies (specifically, the energies involved in processes in LHC) satisfy the condition $E \ll E_p$. Thus, the well-known local QFT in continuous space-time [60]–[63] is verified experimentally only at low energies $E \ll E_p$.

Another scale of energies E is associated with Planck's energy range $E \approx E_p$, where there are no direct experiments and which draws different approaches to quantum-gravitational studies (for example, [64]). Consequently, of great importance is a study of high-energy physics in the "intermediate energy range" $\widetilde{E}$:$(E \ll E_p) < \widetilde{E} < E_p$,which is still unattainable for experiments. The first studies of this range were performed by the author in [57]–[59], in [65] a very important result was demonstrated, and in this paper the author continues his studies for a quantum theory satisfying GUP.

It is obvious that, for $|N_i| \gg 1, |N_t| \gg 1$, the suggested discrete deformation is very close to the initial continuous theory. Besides, the "intermediate energy range" $\widetilde{E}$ that is far from the Planck energies is still not studied experimentally. Because of this, the QFT applicability boundary is of great interest. In the following Section this boundary is determined with the help of black hole thermodynamics for a very significant particular case.

5. The Quantum Black Holes and QFT

The present section is based on the results introduced in [65].

It is supposed that a large (i. e, classical) four-dimensional Schwarzschild black hole is existent with the metric

$$ds^2 = \left(1 - \frac{2MG}{r}\right) dt^2 - \left(1 - \frac{2M}{r}\right)^{-1} dr^2 - r^2 d\Omega^2, \quad (74)$$

where M is the mass of this black hole, and the Schwarzschild horizon radius r_{BH} is defined by

$$r_{BH} = 2MG.. \quad (75)$$

As shown in [66],[67], EP is violated for an observer distant from the black hole event horizon. Considering our objective, it seems expedient to give in brief the main results from [66],[67]. In view of the Unruh effect, an accelerating observer detects thermal radiation (so-called Unruh radiation) with the Unruh temperature given by [68]

$$T_{Unruh} = \frac{\hbar a}{2\pi}, \quad (76)$$

where $a = |\mathbf{a}|$ is a corresponding acceleration.

When an observer is at the *fixed* distance, $r > r_{BH}$, from a Schwarzschild black hole of the mass M and with the event horizon radius $r_{BH} = 2GM$, then, due to the existence of Hawking radiation [69], the observer will measure radiation with thermal spectrum and a temperature given by formula [66]

$$T_{H,r} = \frac{\hbar}{8\pi GM\sqrt{1 - \frac{r_{BH}}{r}}}, \quad (77)$$

where $r > r_{BH}$.Here the normalization $c = k_B = 1$ is used.

To detect Unruh radiation, one can use an Unruh-Dewitt detector [70]–[72]. This detector is an idealized point particle with the internal energy levels labeled by the energy E, coupled via the monopole interaction with a scalar field ϕ. The detector moves along the world line described by the functions $x^{\mu}(\tau)$, where τ is the detector proper time. The "detector-field" interaction is described by the interaction Lagrangian $cm(\tau)\phi[x(\tau)]$, where c is a small coupling constant and m is the detector's monopole moment operator. Suppose the field ϕ is in the vacuum state $|0_M>$, where $|0_M>$ is the "Minkowski vacuum" [70].

For the general trajectory, the detector will not remain in its ground state E_0, but will undergo a transition to the excited state $E > E_0$, while the field will make a transition to the excited state $|\psi>$. For sufficiently small c, the amplitude at this transition may be given by the first-order perturbation theory as [70]

$$ic\langle E,\psi|\int_{-\infty}^{\infty} m(\tau)\phi[x(\tau)]d\tau|0_M,E_0\rangle. \tag{78}$$

The transition amplitude (78), in virtue of the time evolution of $m(\tau)$

$$m(\tau)=e^{H_0\tau}m(0)e^{-H_0\tau},$$

may be transformed (factorized) as

$$ic\langle E|m(0)|E_0\rangle\psi|\int_{-\infty}^{\infty} e^{(E-E_0)\tau}\langle\psi|\phi(x)|0_M\rangle d\tau. \tag{79}$$

Here $H_0|E\rangle=E|E\rangle$ [70].

Then, squaring the modulus (79) and summing over E and over the complete set ψ, one can calculate the transition probability for all possible E and ψ:

$$c^2\sum_E|\langle E|m(0)|E_0\rangle|^2\mathcal{F}(E-E_0), \tag{80}$$

where $\mathcal{F}(E)$ is the detector *response function*

$$\mathcal{F}(E)=\int_{-\infty}^{\infty} d\tau\int_{-\infty}^{\infty} d\tau' e^{-E(\tau-\tau')}G^{+}(x(\tau),x(\tau')). \tag{81}$$

The detector response function $\mathcal{F}(E)$ is independent of its details, is determined by the positive frequency Wightman Green function G^+ and represents a bath of the "particles" effectively experienced by the detector as a result of its motion [70].

Within the scope of the Unruh effect, a number of the above-mentioned particles (Rindler particles) in the Minkowski vacuum comes to $|0_M>$ – this is a thermal Planck spectrum of Rindler particles with temperature given by formula (76).

In [66] it is shown that an observer, positioned at the fixed distance $r > r_{BH}$ from the above-mentioned black holes and measuring Hawking temperature with the value $T_{H,r}$, experiences the local acceleration

$$a_{BH,r} = \frac{1}{\sqrt{1-\frac{r_{BH}}{r}}}\left(\frac{r_{BH}}{2r^2}\right) . \tag{82}$$

Another observer in the Einstein elevator, moving with acceleration through Minkowski space-time, will measure the same acceleration toward the floor of the elevator, thermal radiation with the Unruh temperature given by formula (76). As shown in [66], a_{BH} is coincident with the quantity a from the formula in (76). Then, substituting the acceleration $a = a_{BH}$ from formula (82) into formula (76), we can obtain a formula for $T_{Unruh,r}$ in this case [67]:

$$T_{Unruh,r} = \frac{\hbar}{2\pi\sqrt{1-\frac{r_{BH}}{r}}}\left(\frac{r_{BH}}{2r^2}\right) . \tag{83}$$

If EP is valid, the quantities $T_{Unruh,r}$ from formula (83) and $T_{H,r}$ in (77) should be coincident for $r > r_{BH}$ to a high degree of accuracy. However, we see that this is not true. In [67], e.g. for $r = 4GM = 2r_{BH}$, we have $T_{H,r} = 4T_{Unruh,r}$.

So, far from the event horizon, EP is not the case. Moreover, violation of EP is the greater, the farther it is from the black hole event horizon. Indeed, for an observer at the distance $r > r_{BH}$ we can write $r = \alpha(r)GM = \frac{1}{2}\alpha(r)r_{BH}, \alpha(r) > 2$. Then

$$T_{Unruh,r} = \frac{\hbar}{2\pi\alpha^2(r)GM\sqrt{1-\frac{2}{\alpha(r)}}}. \tag{84}$$

In this way $T_H/T_{Unruh,r} = \alpha^2(r)/4$. And the ratio is the greater, the higher $\alpha(r)$, i.e., the farther from horizon the observer is. Next, for compactness, we denote $T_{Unruh,r}$ in terms of $T_{U,r}$. Of course, in this case we bear in mind only an observer at a sufficiently great but finite distance from the black hole, i.e., only when a gravitational field is thought significant and must be taken into consideration. Thus, in the general case $r_{BH} < r \ll \infty$, whereas in the case of a distant observer we have

$$r_{BH} \ll r \ll \infty. \tag{85}$$

As noted in [65], at Planck's scale EP becomes invalid because at such scales the smooth space-time structure is disturbed. But, since the well-known QFT is a local theory valid in a flat space [60]–[63], it becomes invalid at Planck's scale.

Main Hypothesis [59],[65].
It is assumed that in the general case EP, and consequently, QFT is valid for the locally smooth space-time only if all the energies E of the particles satisfy the necessary condition

$$E \ll E_p. \tag{86}$$

It should be noted that, provided the hypothesis is true, it should be true no matter whether QFT is studied in the perturbative or non-perturbative mode, because the local geometry of space-time is not dependent on this fact.

In [65] the **Main Hypothesis** has been proven in the assumption that space-time foam consists of micro black holes (**mbh**), with the event horizon radius $r \approx l_p$ and of the mass $m \approx m_p$. Note that a model for the **quantum foam** comprising micro black holes (mbh) is most probable from the viewpoint of a correct passage to low energies [73],[74],[75].

Obviously, the case of violated EP in [66],[67] is not directly associated with the **Main Hypothesis** concerning the boundaries of EP validity (formula (86)), because in [66],[67] consideration is given to a large black hole with the event horizon radius r_{BH} much greater than the Planck length $r_{BH} \gg l_p$ at sufficiently low energies.

The principal result from the remarkable papers [66],[67] may be summarized as follows:

Comment 5.1. In any point of space-time that is in a field of a large classical Schwarzschild black hole, and in the cases when this field must be taken into consideration, it is impossible to remove this field in the vicinity of the point even locally, i.e., to consider space-time as flat.

Comment 5.2. It is important to refine some formulations from [66],[67]. Specifically, if $r \rightarrow r_{BH}$, then $T_{H,r} \rightarrow \infty, T_{Unruh,r} \rightarrow \infty$ in formulae (77) and (83), respectively. Note that for $r \rightarrow r_{BH}$ these temperatures become infinite $T_{H,r} = \infty, T_{Unruh,r} = \infty$. Based on this fact, in [66],[67] it is inferred "that the equivalence principle is restored on the horizon". But this statement is not correct. Restoration of EP is not following from the fact that the above temperatures take infinite values. We can only state that temperature on the BH horizon and in its vicinity cannot be the parameter detecting a deviation from EP. In the opposite case one can arrive at violation: on the black hole event horizon, where a gravitational field is very large in value, EP holds, whereas far from the event horizon, where a gravitational field is much weaker, this principle is violated.

The question arises: what changes in the foregoing pattern are associated with the substitution of **mbh** for a large four-dimensional Schwarzschild black hole? Note that **mbh** has Planck's parameters, i.e., we have

$$r_{mbh} \propto l_p, m_{mbh} \propto m_p. \tag{87}$$

However, **mbh** is not an *extremal black hole*, i.e., **mbh** with minimal values of $r_{mbh} = r_{min}, m_{mbh} = r_{min}$, as the latter has no Hawking radiation [76]. Without loss of generality, it is assumed that the above-mentioned **mbh** is sufficiently close to *extremal black hole*:

$$r_{mbh} \approx r_{min}, m_{mbh} \approx m_{min}, \tag{88}$$

but is not coincident with it $r_{mbh} \neq r_{min}, m_{mbh} \neq m_{min}$.

According to the present-day knowledge, a semi-classical region starts between 5 and 20 times the Planck scale [77]. Then, despite the fact that this **mbh** is itself in the quantum gravity region, the scales of $r \gg r_{mbh}$ determine the region, where a semi-classical approximation is valid.

Then it follows that, for the energies E somewhat lower than the Planck energies (i.e., $E \ll E_p$) involved in the condition (86), a semi-classical approximation is valid. This means that results of [66],[67] are valid for **mbh** with the corresponding quantum corrections [64]. These quantum corrections in the four-dimensional Schwarzschild classical black hole case are vanishingly small. Nevertheless, for **mbh** they are already high.

In [65] it is demonstrated that, if $r \gg r_{mbh}$ (semi-classical approach), then $T_{H,r,q} \gg T_{U,r,q}$,where $T_{H,r,q}$ is the temperature of a Schwarzschild **mbh** but with regard to the corresponding quantum corrections (for example, [64],[78],[79],[80]). Similarly, $T_{U,r,q}$ is the Unruh temperature with the corresponding quantum corrections. So, in this case for **mbh** there is an analog of **Comment 5.1**.

Comment 5.1*. In any point of space-time that is in a field of a Schwarzschild **mbh**, and in the cases when this field must be taken into consideration, it is impossible to remove this field in the vicinity of the point even locally, i.e., to consider space-time as flat.

As canonical QFT is the local theory valid in space-time with the flat geometry [62],[63], this suggests a proof for the **Main Hypothesis** assumption that quantum foam consists of micro black holes (**mbh**).

In such a manner in this case the canonical QFT has some ultraviolet limit (boundary application) lying considerably below the Planck scales of the natural ultraviolet cutoff point that is at the energies E significantly lower than the Planck's $E \ll E_p$.

Comment 5.3. In the case of **mbh** **Comment 5.2** is absolutely clear. In fact, at a horizon of **mbh**, i.e., for $r = r_{mbh}$, $T_{H,r} = T_{Unruh,r} = \infty$ similar to large black holes but , naturally, without any restoration of EP as the domain $r = r_{mbh} \propto l_p$ is the region of Planck energies or of quantum foam, where EP in its canonical formulation becomes invalid. It is obvious that at the event horizon $r = r_{mbh}$ of **mbh** and in its vicinity a gravitational field becomes very strong due to quantum effects, and nothing could destroy it.

6. Conclusion

6.1. Based on the above results, all the energies E we can classify into 3 groups [65]:

a) low energies $0 < E \leq \widetilde{E} \ll E_p$ – energies, for which the Strong Equivalence Principle is valid in virtue of formula (86), and hence this energy interval sets the QFT applicability boundaries.
a1) Since $\widetilde{E} \ll E_p$, it is natural to assume that $\widetilde{E} \approx 10^{-N}E_p$, where $N \geq 2$. Obtaining of more accurate estimates for N is a separate problem;

b) intermediate energies $\widetilde{E} < E < E_p$ – energies, for which the Strong Equivalence Principle and, consequently QFT, becomes invalid but the corresponding scales are greater than the Planck. It can be assumed that QFT in this energy range will be a theory in a gravitational field that could not be destroyed even locally. In the case under study it is assumed that this field is created by **mbh**. Impossibility of destroying this field even locally is associated with large quantum corrections for the corresponding quantities which should be taken into consideration at these energies [78],[79],[80].
Let us identify the energy scale $\widetilde{E} < E < E_p$ as *prequantum gravity phase*;

c) high (essentially maximal) energies $E \approx E_p$ or $E > E_p$. This interval is the region of quantum gravity energies.

6.2. It is absolutely clear that, though the material in Section 5 and the results in it are associated with the canonical QFT in continuous space-time, they are still valid for the discrete variant of a theory given in Section 2. In [58],[59] a quantum theory is considered only for the energies satisfying the conditions $E \ll E_p$ and $E \approx E_p$ rather than for a complete scale of energies. In [65] and in the present paper this gap is eliminated.

6.3. Returning to Section 5, we should note an important difference between a semi-classical consideration in [66],[67] and the results in [65] or in this paper. In the first case a **UDW** detector may be without limit close to the classical BH event horizon, because in this case the semi-classical approximation is still valid. But in the case under study a **UDW** detector should be sufficiently distant from the **mbh** event horizon, otherwise it can fall into the quantum-gravity region and become meaningless.

6.4. The subject-matter of this work is associated with such a problem as experimental detection of a quantum foam. At the present time, the relevant results are still inconsistent [81] due to inadequate capacity of the data available and insufficient precision of the experiments. (The situation resembles that in chronology of the detection of gravitational waves the relict including.) Nevertheless, an active search in this direction is in progress [82], [83].

6.5. Finally, it is important to note that Albert Einstein formulated his strong Equivalence Principle only for classical objects against a classical space-time background. This is explained by the fact that in the period, when General Relativity was created (1915,1916) and this principle was formulated, a quantum theory was at the stage of conception and its main postulates were unknown. Because of this, formulation of the principal result from [66],[67] is not quite exact. Strong EP, in its initial formulation, is still valid and the principal result from [66],[67] should be as follows:
generalization of the strong Equivalence Principle for the case of a semi-classical approximation in a black hole gravitational field becomes invalid.

Thus, the main problem of QFT applicability in the perturbative or nonperturbative mode is associated with definition of the scale $\widetilde{E}$ in point **6.1**.

Conflict of Interests

The author declares that there is no conflict of interests regarding the publication of this paper.

References

[1] Shalyt-Margolin A.E., Minimal Length and the Existence of Some Infinitesimal Quantities in Quantum Theory and Gravity, *Adv. High Energy Phys.,* **2014** (2014), 8.
http://dx.doi.org/10.1155/2014/195157.

[2] Shalyt-Margolin Alexander, Minimal Length, Measurability and Gravity, *Entropy,* **18(3)** (2016), 80. http://dx.doi.org/10.3390/e18030080.

[3] Shalyt-Margolin Alexander, Minimal Length at All Energy Scales and Measurability, *Nonlinear Phenomena in Complex Systems,* **19(1)** (2016), 30–40.

[4] Shalyt-Margolin A. E., Uncertainty Principle at All Energies Scales and Measurability Conception for Quantum Theory and Gravity, *Nonlinear Phenomena in Complex Systems,* **19(2)** (2016), 166–181.

[5] Shalyt-Margolin Alexander E., Minimal Length, Minimal Inverse Temperature, Measurability and Black Hole,*Electronic Journal of Theoretical Physics,***14(37)** (2018), 35-54.

[6] Shalyt-Margolin Alexander, The Measurability Notion in Quantum Theory, Gravity and Thermodynamics: Basic Facts and Implications, Chapter Eight in *Horizons in World Physics. Volume 292*, Reimer, A., Ed.; Nova Science: Hauppauge, NY, USA, 2017; 199–244.

[7] Shalyt-Margolin A.E. and Tregubovich A.Ya., *Mod. Phys. Lett. A*, **19**, 71(2004).

[8] Heisenberg W., *Z. Phys.* **43**, 172(1927). (In German).

[9] Bohr N., *Faraday Lectures* pp. 349-384, 376-377 Chemical Society, London (1932).

[10] Heisenberg W., *Der Teil und Das Ganze* [*The part and the whole*] ch 9 R. Piper, Munchen, 1969.

[11] Lindhard J., Complementarity between energy and temperature. In: *The Lesson of Quantum Theory*; Ed. by de Boer J., Dal E. and Ulfbeck O. North-Holland, Amsterdam 1986.

[12] Lavenda B., *Statistical Physics: A Probabilistic Approach*. J. Wiley and Sons, N.Y., 1991.

[13] Mandelbrot B., *IRE Trans. Inform. Theory*, **IT-2**, 190 (1956).

[14] Rosenfeld L. In *Ergodic theories*; Ed. by P. Caldrirola Academic Press, N. Y., 1961.

[15] Schlogl F., *J. Phys. Chem. Solids*, **49**, 679 (1988).

[16] Uffink J., Lith-van Dis J., *Found. of Phys.* **29**, 655 (1999).

[17] Veneziano G. A., *Europhys. Lett.*, **2**, 199 (1986).

[18] Amati D., Ciafaloni M. and Veneziano G. A., *Phys. Lett. B*, **216**, 41 (1989).

[19] Witten E., *Phys. Today*, **49**, 24 (1996).

[20] Adler R. J. and Santiago D. I., *Mod. Phys. Lett. A*, **14**, 1371 (1999).

[21] Ahluwalia D. V., *Phys. Lett. A*, **A275**, 31 (2000).

[22] Ahluwalia D. V., *Mod. Phys. Lett. A*, **A17**, 1135 (2002).

[23] Maggiore M., *Phys. Lett. B.*, **319**, 83 (1993).

[24] Maggiore M., *Phys. Rev. D.*, **49**, 2918 (1994).

[25] Maggiore M., *Phys. Rev. D.*, **48**, 65 (1993).

[26] Capozziello S., Lambiase G. and Scarpetta G., *Int. J. Theor. Phys.*, **39**, 15 (2000).

[27] Kempf A., Mangano G. and Mann R. B., *Phys. Rev. D.*, **52**, 1108 (1995).

[28] Nozari K. and Etemadi A., *Phys. Rev. D.*, **85**, 104029 (2012).

[29] Farmany A., *Acta Phys. Pol. B.*, **40**, 1569 (2009).

[30] Carroll R., Fluctuations, Information, Gravity and the Quantum Potential. *Fundam. Theor. Phys.* 148, Springer, N. Y., 2006; 454pp.

[31] Hassan S. F. and Martin Sloth M. S., *Nucl. Phys. B.*, **674**, 434 (2003).

[32] Berestetskii V. B., Lifshitz E. M., Pitaevskii L. P., *Relativistic Quantum Theory*, Pergamon, Oxford, UK, 1971.

[33] Shalyt-Margolin A. E., Suarez J.G., *Int. J. Mod. Phys. D.*, **12**, 1265 (2003).

[34] Shalyt-Margolin A. E., *Mod. Phys. Lett. A.*, **19**, 391 (2004).

[35] Shalyt-Margolin A. E., *Mod. Phys. Lett. A.*, **19**, 2037 (2004).

[36] Shalyt-Margolin A. E., *Int. J. Mod. Phys. D.*, **13**, 853 (2004).

[37] Shalyt-Margolin A. E., *Int. J. Mod. Phys. A.*, **20**, 4951(2005).

[38] Faddeev L., *Priroda*, **5**, 11 (1989).

[39] Hooft Gerard't, Giddings Steven B., Rovelli Carlo, Nicolini Piero, Mureika Jonas, Kaminski Matthias, and Bleicher Marcus, The Good, the Bad, and the Ugly of Gravity and Information, [arXiv:1609.01725v1 [hep-th] 6 Sep 2016].

[40] Padmanabhan T., *Class. Quant. Grav.*, **19**, 5387 (2002).

[41] Padmanabhan T., *International Journal of Modern Physics D.*, **14**, 2263 (2005).

[42] Padmanabhan T., *General Relativity and Gravitation*, **34**, 2029 (2002).

[43] Padmanabhan T., *Brazilian Journal of Physics*, **35**, 362 (2005).

[44] Padmanabhan T., *International Journal of Modern Physics D.*, **15**, 1659 (2006).

[45] Mukhopadhyay A. and Padmanabhan T., *Physical Review D.*, **74**, ID 124023 (2006).

[46] Padmanabhan T., *General Relativity and Gravitation*, **40**, 529 (2008).

[47] Padmanabhan T. and Paranjape A., *Physical Review D*, **75**, ID 064004 (2007).

[48] Padmanabhan T., AIP Conference Proceedings, **939**, 114 (2007).

[49] Padmanabhan T., *Physics Reports*, **406**, 49 (2005).

[50] Paranjape A., Sarkar S., and Padmanabhan T., *Physical Review D*, **74**, ID 104015 (2006).

[51] Padmanabhan T., *Reports on Progress in Physics*, **74**, ID 046901 (2010).

[52] Shalyt-Margolin A. E., *International Journal of Modern Physics D*, **21**, ID 1250013 (2012).

[53] Frolov Valery P. and Novikov Igor D., *Black Hole Physics: Basic Concepts and New Developments*, Springer-Verlag, 1998.

[54] Kobayashi S., Nomizu K., *Foundations of Differential Geometry*, V. II, Interscience Publishers, New York-London-Sydney 1969.

[55] Tawfik, Abdel Nasser; Diab, Abdel Magied, Generalized Uncertainty Principle: Approaches and Applications. *Int. J. Mod. Phys. D.* **2014**, *23*, 1430025.

[56] Messiah, A., *Quantum Mechanics*, North Holland Publishing Company, Amsterdam, The Netherlands, Volumes 1, 2, 1961.

[57] Shalyt-Margolin, A., The Equivalence Principle Applicability Boundaries, QFT in Flat Space and Measurability I. Free Quantum Fields. *Int. J. Nonlinear Phenomena in Complex Systems.* **2019**, *22(2)*, 135 - 150.

[58] Shalyt-Margolin, A., QFT in Flat Space and Measurability II. Perturbation Theory for a Scalar Field Model. *Int. J. Nonlinear Phenomena in Complex Systems.* **2020**, *23(1)*, 33 - 53.

[59] Shalyt-Margolin, A., The Equivalence Principle Applicability Boundaries, Measurability, and UVD in QFT. *Int. J. Nonlinear Phenomena in Complex Systems.* **2021**, *24(1)*, 38 - 55.

[60] Rayder Lewis H., *Quantum Field Theory*, University of Kent and Canterbuty.

[61] Cheng Ta-Pei, Li Ling-Fong, *Gauge Theory of elementary particle physics*, Oxford Scince Publications, 548 pages.

[62] Peskin M.E., Schroeder D.V., *An Introduction to Quantum Field Theory*, Addison-Wesley Publishing Company, 1995.

[63] Weinberg Steven, *The Quantum Theory of Fields*, Vol. 1, 2. Cambridge University Press, 1995.

[64] Kiefer Claus, *Quantum Gravity.Third Edition*, Oxford University Press, United Kingdom (2012).

[65] Shalyt-Margolin A., The Quantum Field Theory Boundaries Applicability and Black Holes Thermodynamics. *Int. J. Theor. Phys.* (2021). https://doi.org/10.1007/s10773-021-04804-1.

[66] Singleton D. and Wilburn S., Hawking Radiation, Unruh Radiation, and the Equivalence Principle. *Phys. Rev. Lett.* **107**, 081102 (2011); *Phys. Rev. Lett.* **108**, 049002 (2012).

[67] Singleton Douglas and Wilburn Steve, Global versus local–Mach principle versus the equivalence principle. *Int. J. Mod. Phys. D.* **25**, 1644009 (2016).

[68] Unruh W. G., Notes on black-hole evaporation. *Phys. Rev. D.* **14**, 870 (1976).

[69] Hawking S. W., Particle Creation by Black Holes. *Comm. Math. Phys.* **43**,199 (1975).

[70] Birrell N. D. and Davies P. C. W., *Quantum fields in curved space*, Cambridge University Press, Cambridge (1982).

[71] De Witt B. S., Quantum gravity: new synthesis, in *General Relativity*, Eds. S.Hawking, W. Israel, Cambridge University Press, 1979.

[72] Akhmedov E.T. and Singleton D., On the relation between Unruh and Sokolov–Ternov effects. *Int. J. Mod. Phys. A.* **22**, 4797 (2007).

[73] Scardigli Fabio, Black Hole Entropy: a spacetime foam approach. *Class.Quant.Grav.* **18**, 1781–1793 (1997).

[74] Scardigli Fabio, Generalized Uncertainty Principle in Quantum Gravity from Micro-Black Hole Gedanken Experiment. *Phys. Lett. B.* **452**, 39–44 (1999).

[75] Scardigli Fabio, Gravity coupling from micro-black holes. *Nucl. Phys. Proc. Suppl.* **88**, 291–294 (2000).

[76] Carroll S. M., Johnson M. C. and Randall L., Extremal limits and black hole entropy. *Journal of High Energy Physics*. **11**, 109 (2009). doi:10.1088/1126-6708/2009/11/109.

[77] Meade P. and Randall L., Black holes and quantum gravity at the LHC. *Journal of High Energy Physics*. **05**, 003 (2008). doi:10.1088/1126-6708/2008/05/003.

[78] Nouicer K., Quantum-corrected black hole thermodynamics to all orders in the Planck length. *Phys. Lett B.* **646**, 63 (2007).

[79] Jamil Mubasher, Darabi Farhad, Quantum corrections for a black hole in an asymptotically safe gravity with higher derivatives. *Int. J. Theor. Phys.* **51**, 1007-1012 (2012).

[80] Gorji M. A., Nozari Kourosh, Vakili B., Polymeric Quantization and Black Hole Thermodynamics. *Phys. Lett. B.* **735**, 1007-1012.

[81] *Integral challenges physics beyond Einstein / Space Science / Our Activities / ESA*.

[82] Moyer, Michael (17 January 2012), "Is Space Digital?". *Scientific American*. Retrieved 3 February 2013.

[83] Cowen, Ron (22 November 2012), "Single photon could detect quantum-scale black holes". *Nature News*. Retrieved 3 February 2013.

Chapter 3

Thermodynamics of Black Hole and Thermal Fluctuations

Sudhaker Upadhyay*
Department of Physics, K. L. S. College, Nawada, Bihar, India
Department of Physics, Magadh University, Bodh Gaya, Bihar, India

Abstract

In this chapter, we first emphasize the thermodynamics and stability of charged quasitopological black holes in equilibrium. Furthermore, we study the effect of small thermal fluctuation around the equilibrium on the thermodynamics of these black holes. Entropy gets corrected with an additional (logarithmic) term due to thermal fluctuation and, therefore, other thermal quantities also get corrected. The impact of such thermal fluctuation on the thermodynamics of black holes is also analyzed comparatively.

1. Introduction

Black holes are some of the most fascinating objects in the universe. The existence of black holes is confirmed now. Black holes cannot be seen directly, but their impacts on surroundings have been established. The gravity of black holes is so strong that nothing, not even light, can escape from that region of

*Corresponding Author's Email: sudhakerupadhyay@gmail.com; sudhaker@associates.iucaa.in.

In: A Guide to Black Holes
Editor: Kenath Arun
ISBN: 979-8-88697-163-7

space-time. A real black hole system can have an accretion disk and jet, which make them quite complicated objects. In contrast, from a mathematical point of view, stationary black holes are vacuum solutions to Einstein's equations which are very simple objects. The reason behind the simplicity is the no-hair theorem [1] which says that black holes are uniquely defined in terms of mass, electric charge, and angular momentum only. Such simple objects are so interesting because their physics merges three different branches of physics: general relativity, quantum mechanics, and statistical physics [2].

Black holes are those solutions to Einstein's gravity that have a real singularity at the center. The physics of black holes at the quantum level is much more complicated and to a certain extent much more interesting. Hawking found that black holes are not truly black but emit a radiation like black body [3]. This discovery has several fascinating consequences. Hawking radiations are not quantum gravity effects but other quantum mechanical effects. Hawking effects imply that a black hole has a temperature and thus an entropy. This is a beautiful result.

The black hole thermodynamics begins with the papers of Bekenstein [4,5] and Hawking [6] and it is found that entropy is related to the area of the horizon. Presently, it is more or less certain that large black holes have an entropy proportional to their horizon area [7–10]. Thus, an interesting question may arise what does happen when one reduces the size of the black holes. Many people attempted to answer this question [11–17]. It is found that the small black holes get influenced by small statistical fluctuations around the equilibrium. This leads to a correction in the area-law by additional terms. This is confirmed that the leading-order correction in the area-law due to small thermal fluctuations around equilibrium are always logarithmic. Thermodynamics of small black holes under the effect of small thermal fluctuations are studied diversely [18–24].

In this chapter, we first review the basics of black hole thermodynamics in Sec. 2. The thermodynamics and stability of a particular black hole, namely charged quasitopological black hole, are discussed in Sec. 3. The effects of small thermal fluctuation on the thermodynamics of small charged quasitopological black holes are presented in Sec. 4. Finally, the results of this chapter are summarized in the last section.

2. Black Hole Thermodynamics

The black holes provide the first signature to develop a relation between gravitation, thermodynamics, and quantum theory. The leading-order contribution to the entropy of a black hole is given by the Bekenstein-Hawking entropy as follows:

$$S_0 = \frac{A}{4}, \tag{1}$$

where A is the area of a $(d-2)$-dimensional slice of the horizon. For simplicity, we set the unit such that the constants G, c, k_B, and $\hbar$ are equal to unity.

Hawking (in 1974) found that the (physical) temperature of a black hole is not absolute zero. So, due quantum particle creation effects, a black hole radiates at temperature

$$T_H = \frac{\kappa}{2\pi} = \left.\frac{f'(r)}{4\pi}\right|_{r=r_+}, \tag{2}$$

where κ is surface gravity and f' is the derivative of metric function.

The first law of black holes [25] reads

$$dM = T_H dS_0 + \Omega dJ + \Phi dQ + PdV, \tag{3}$$

where J and Q are the angular momentum and charge of the black hole, respectively and their respective thermodynamic conjugates are the angular velocity Ω and the electromagnetic potential Φ. The the last term corresponds to the familiar pressure-volume term from thermodynamics, with thermodynamic pressure P and conjugate thermodynamic volume V.

The correspondence with black hole thermodynamics with standard thermodynamics can be listed as follows:

Standard Thermodynamics	Black Hole Thermodynamics
Enthalpy	Mass
Temperature	Surface Gravity
Entropy	Horizon Area
Pressure	Cosmological Constant

The realization of charged black holes as Van der Waals fluids was one of the first implications of the black hole thermodynamics [26]. It is well-known that the Van der Waals approximates the the behavior of real fluids by modifying the equation of state for an ideal gas by considering the attraction between the molecules and their finite size parameterized by the constants a and b, respectively as follows:

$$P = \frac{T}{v-b} - \frac{a}{v^2}. \tag{4}$$

Here, v is the specific volume of the fluid. Quite surprisingly, it is found that charged AdS black holes follows the same basic relationships.

3. Equilibrium Thermodynamics of Charged Quasitopological Black Holes

In this section, we implement the above thermodynamical set for the specific black hole. For this purpose, we write a general form of the action for the (quasitopological) gravity in $(d+1)$ space-time dimensions in presence of the electromagnetic field as follows [27,28]

$$\begin{aligned} \mathcal{A} &= \frac{1}{16\pi}\int d^{d+1}x\sqrt{-g}\left[R - \Lambda + \frac{\lambda l^2}{(d-2)(d-3)}\mathcal{Y}_1\right. \\ &+ \left.\frac{8(2d-1)\mu l^4}{(d-2)(d-5)(3d^2-9d+4)}\mathcal{Y}_2 - \frac{1}{4}F_{ab}F^{ab}\right]. \end{aligned} \tag{5}$$

Here, the cosmological constant can be defined in terms of Planck's length l as $\Lambda = -d(d-1)/2l^2$. However, $F_{ab} = \partial_a A_b - \partial_b A_a$ is Maxwell field-strength tensor, λ is the Gauss-Bonnet coupling constant and μ is the quasitopological coupling constant. Here, $\mathcal{Y}_1$ and $\mathcal{Y}_2$ are the Gauss-Bonnet and quasitopological terms, respectively, with following explicit expressions:

$$\mathcal{Y}_1 = \mathcal{R}_{abcd}\mathcal{R}^{abcd} - 4\mathcal{R}_{ab}\mathcal{R}^{ab} + \mathcal{R}^2, \tag{6}$$

$$\begin{aligned} \mathcal{Y}_2 &= \mathcal{R}_{a\ b}^{\ c\ d}\mathcal{R}_{c\ d}^{\ e\ f}\mathcal{R}_{e\ f}^{\ a\ b} + \frac{1}{(2d-1)(d-3)}\left[\frac{3(3d-5)}{8}\mathcal{R}_{abcd}\mathcal{R}^{abcd}\mathcal{R}\right. \\ &- 3(d-1)\mathcal{R}_{abcd}\mathcal{R}^{abc}{}_{e}\mathcal{R}^{de} + 3(d+1)\mathcal{R}_{abcd}\mathcal{R}^{ac}\mathcal{R}^{bd} \qquad (7) \\ &+ \left. 6(d-1)\mathcal{R}_a{}^b\mathcal{R}_b{}^c\mathcal{R}_c{}^a - \frac{3(3d-1)}{2}\mathcal{R}_a{}^b\mathcal{R}_b{}^a\mathcal{R} + \frac{3(d+1)}{8}\mathcal{R}^3\right]. \qquad (8) \end{aligned}$$

Now, in order to discuss thermodynamics of this black hole described by the action (5), we write $(d+1)$-dimensional static metric as following:

$$ds^2 = N^2(r)f(r)dt^2 + \frac{dr^2}{f(r)} + r^2 \sum_{i=1}^{d-1} d\phi_i^2, \tag{9}$$

where $N(r)$ is a lapse function and $\sum_{i=1}^{d-1} d\phi_i^2$ is the metric on a compact $d-1$-dimensional space.

Dehghani and Vahidinia in Ref. [28] found that $N(r)$ must be a constant and (without the loss of generality) can be set to unity.

For the gauge potential $A_a = h(r)\delta_a^t$, the equation of motion has following solution:

$$h(r) = -\sqrt{\frac{2(d-1)}{d-2}}\frac{q}{r^{d-2}}.$$

Now, the metric function is calculated by [28]

$$f(r) = \frac{r^2}{l^2} - \frac{m}{r^{d-2}} + \frac{q^2}{r^{2(d-2)}}. \tag{10}$$

Here, q and m are integration constants related to the electric charge and mass of the black holes, respectively. The mass parameter m can easily be determined from the vanishing metric function on the horizon

$$f(r)|_{r=r_+} = 0.$$

This leads to

$$m = \frac{r_+^d}{l^2} + \frac{q^2}{r_+^{d-2}}. \tag{11}$$

Now, relation (2) for above metric function (10) gives the Hawking temperature of the event horizon [28]

$$T_H = \frac{dr_+ - (d-2)q^2 l^2 r^{3-2d}}{4\pi l^2}, \tag{12}$$

where r_+ is the outer horizon radius. The potential, measured at infinity with respect to the horizon, can be given by [29]

$$\Phi = A_a\xi^a|_{r\to\infty} - A_a\xi^a|_{r=r_+}, \tag{13}$$

where ξ^a refers to the null generator of the horizon. The electric potential for static case ($\xi^a = \delta_t^a$) is given by [4, 28, 30, 31]

$$\Phi = \sqrt{\frac{2(d-1)}{d-2}}\frac{q}{r_+^{d-2}}. \tag{14}$$

The Gibbs free energy G identifies as the Euclidean action $\mathcal{A}_E$ times the Hawking temperature T_H [32]. The Euclidean action is calculated as

$$\mathcal{A}_E = -\frac{1}{16\pi l^2 T_H^{-1}}\left(r_+^d + \frac{q^2 l^2}{r_+^{d-2}}\right). \tag{15}$$

This gives Gibbs free energy as

$$G = \mathcal{A}_E T_H = -\frac{1}{16\pi l^2}\left(r_+^d + \frac{q^2 l^2}{r_+^{d-2}}\right).$$

In terms of T_H and Φ, Gibbs free energy is given by

$$G(T_H, \Phi) = -\frac{1}{16\pi l^2}\left(r_+^d + \frac{d-2}{2(d-1)} l^2 \Phi^2 r_+^{d-2}\right). \tag{16}$$

Exploiting relations (12) and (14), the Hawking temperature can be expressed in terms of electric potential Φ as

$$T_H = \frac{d r_+}{4\pi l^2} - \frac{(d-2)^2 \Phi^2}{8\pi(d-1) r_+}. \tag{17}$$

Consequently, the horizon radius can be expressed in terms of temperature and electric potential as

$$r_+ = \frac{2\pi l^2}{d} T_H + 2\left[\frac{\pi^2 l^4}{d^2} T_H^2 + \frac{(d-2)^2 l^2}{8d(d-1)}\Phi^2\right]^{1/2}. \tag{18}$$

The entropy per volume of charged black (brane) hole is calculated as [28]

$$S_0 = -\left(\frac{\partial G}{\partial T_H}\right)_\Phi = \frac{1}{4} r_+^{d-1}. \tag{19}$$

The charge density of charged black hole is calculated by

$$Q = -\left(\frac{\partial G}{\partial \Phi}\right)_{T_H} = \frac{\sqrt{2(d-1)(d-2)}}{16\pi} q. \tag{20}$$

The energy (total mass) density of charged black hole is calculated by

$$M = G + T_H S_0 + \Phi Q = \frac{d-1}{16\pi} m. \tag{21}$$

We observe that the entropy calculated in Eq. (19) is in complete agreement with the entropy calculated in [27]. Also, the charge density given in (20) is equal to that calculated using the Gauss law.

3.1 Stability of Charged Quasitopological Black Holes in Equilibrium

Here, we shall present the thermal stability of the charged quasitopological black holes in thermal equilibrium. One should remember that the stability of a thermal system in the canonical ensemble is described by the signature of heat capacity. The change in the signature may occur when heat capacity meets root(s) or divergence(es). The root of heat capacity (or temperature) confirms the bound point. This bound point divides physical solutions (describing positive temperature) from non-physical solutions (describing negative temperature). The divergences of heat capacity occur at the phase transition points. The a negative value of heat capacity represents the unstable black holes which may acquire a stable state after phase transition.

The heat capacity density with a fixed electric (chemical) potential, Φ, is given by

$$\begin{aligned} C_\Phi &= T_H \left(\frac{\partial S_0}{\partial T_H} \right)_\Phi, \\ &= \frac{2\pi (d-1)^2 l^2 r_+^d}{2d(d-1) r_+^2 + (d-2)^2 l^2 \Phi^2}, \\ &= \frac{\pi (d-1) l^2 r_+^{d-1}}{d r_+^{2d-2} + (d-2) l^2 q^2 r_+^{3-2d}}. \end{aligned} \tag{22}$$

To check bound points, we solve the denominator of above expression with respect to horizon radius and get

$$r_c = \left[-\frac{(d-2) l^2 q^2}{d} \right]^{1/4d-3}. \tag{23}$$

However, to get phase transition points, one can solve the numerator of above expression concerning horizon radius. From the expression, it is evident that for quasitopological black holes in four space-time dimensions there exists no phase transition point. We also see that thermally equilibrium black holes are stable.

4. Corrected Microcanonical Entropy due to Thermal Fluctuation

In this section, we discuss corrections to thermodynamic entropy of the small black holes when fluctuations around equilibrium are taken into account. In this connection, the system of black holes can be treated as the canonical ensemble. The partition function is defined by

$$Z(\beta) = \int_0^\infty \rho(E) e^{-\beta E} dE, \tag{24}$$

where $\beta = T^{-1}$ as Boltzmann constant is set to unity.

The inverse Laplace transform keeping E fixed gives the density of states [33, 34]

$$\rho(E) = \frac{1}{2\pi i} \int_{\gamma - i\infty}^{\gamma + i\infty} Z(\beta) e^{\beta E} d\beta = \frac{1}{2\pi i} \int_{\gamma - i\infty}^{\gamma + i\infty} e^{\mathcal{S}(\beta)} d\beta, \tag{25}$$

where $\mathcal{S}(\beta)$ is the exact entropy at equilibrium which depends on temperature ($\beta = T^{-1}$) explicitly. The exact entropy describes total sum of entropy of sub-systems of the thermodynamical system, which are small enough to be in equilibrium.

To simplify the complex integral given in (25), one can use the method of steepest descent around the saddle point $\beta_0 (= 1/T_0)$ such that $\left(\frac{\partial \mathcal{S}(\beta)}{\partial \beta}\right)_{\beta = \beta_0} = 0$. The following equilibrium relation holds for the the equilibrium temperature T_0:

$$E = -\left(\frac{\partial Z}{\partial \beta}\right)_{\beta = \beta_0}.$$

Now, exact entropy $\mathcal{S}(\beta)$ around the saddle point $\beta = \beta_0$ can be expanded using the Taylor expansion as

$$\mathcal{S}(\beta) = S_0 + \frac{1}{2}(\beta - \beta_0)^2 \left(\frac{\partial^2 \mathcal{S}(\beta)}{\partial \beta^2}\right)_{\beta = \beta_0} + (\text{higher order terms}). \tag{26}$$

Here, $S_0 := \mathcal{S}(\beta_0)$ refers to the equilibrium entropy. The density of states expressed in (25) corresponding to the expanded form of $\mathcal{S}(\beta)$ is given by

$$\rho(E) = \frac{e^{S_0}}{2\pi i} \int_{\gamma - i\infty}^{\gamma + i\infty} d\beta e^{\frac{1}{2}(\beta-\beta_0)^2 \left(\frac{\partial^2 \mathcal{S}(\beta)}{\partial \beta^2}\right)_{\beta=\beta_0}}. \tag{27}$$

By choosing $\gamma = \beta_0$ for positive $\left(\frac{\partial^2 \mathcal{S}(\beta)}{\partial \beta^2}\right)_{\beta=\beta_0}$ the above integral simplifies to [17]

$$\rho(E) = \frac{e^{S_0}}{\sqrt{2\pi \left(\frac{\partial^2 \mathcal{S}(\beta)}{\partial \beta^2}\right)_{\beta=\beta_0}}}. \tag{28}$$

Since the density of states $\rho(E)$ and $\left(\frac{\partial^2 \mathcal{S}(\beta)}{\partial \beta^2}\right)_{\beta=\beta_0}$ have dimensions of inverse energy and energy squared, respectively. Thus, the logarithm of $\rho(E)$ is the microcanonical entropy and given by

$$S := \ln \rho(E) = S_0 - \frac{1}{2} \log \left(\frac{\partial^2 \mathcal{S}(\beta)}{\partial \beta^2}\right)_{\beta=\beta_0} + (\text{sub-leading terms}). \tag{29}$$

This S refers to the corrected microcanonical entropy at equilibrium, obtained by incorporating small fluctuations around the thermal equilibrium. However, exact entropy, $\mathcal{S}(\beta)$, describes entropy at any temperature (and not just at equilibrium). The above expression (29) holds for all thermodynamic systems considered as a canonical ensemble (including black holes).

A detailed calculation discussed in Ref. [17] suggests that

$$\left(\frac{\partial^2 \mathcal{S}(\beta)}{\partial \beta^2}\right)_{\beta=\beta_0} = T^2 S_0. \tag{30}$$

Thus, the corrected microcanonical entropy at equilibrium by considering small fluctuations around thermal equilibrium is given by

$$S = S_0 - \frac{1}{2} \log[T^2 S_0]. \tag{31}$$

The sub-leading terms are very small and can be neglected.

For the system of black holes, the above expression results to [17]

$$S = S_0 - \frac{1}{2} \alpha \log[T_H^2 S_0], \tag{32}$$

where S_0 and T_H are the Bekenstein-Hawking area law and Hawking temperature, respectively. A parameter α is introduced to track the contribution of thermal fluctuation as $\alpha = 0$ corresponds to no correction due to thermal fluctuation and $\alpha = 1$ corresponds to correction due to thermal fluctuation.

4.1 Effects of Thermal Fluctuations on Thermodynamics of Quasitopological Black Holes

Corresponding to Hawking temperature (12) and Bekenstein-Hawking area law (19) of the charged quasitopological black hole, S given in (32) takes the following form:

$$\begin{aligned} S &= \frac{1}{4}r_+^{d-1} - \frac{1}{2}\alpha \log\Big[d^2 r_+^{d+1} + (d-2)^2 q^4 l^4 r_+^{5-3d} \\ &- 2d(d-2)q^2 l^2 r_+^{3-d}\Big] + \alpha \log[8\pi l^2]. \end{aligned} \tag{33}$$

The effects of small thermal fluctuation around equilibrium on the entropy density can be seen in figures (1) and (2) for $d = 3$ and $d = 4$, respectively. Here, we see that thermal fluctuation plays a significant role for small black holes only. For points like black holes in $d = 3$ due to fluctuations, the system has a negative asymptotic value. As long as the horizon radius increases the entropy increases abruptly and has a peak and after a certain size of black hole entropy again starts falling. Corrected entropy depends on Plank's length and charge also as described in the figure. For $d = 4$, corresponding to specific charge and length the corrected entropy is not a continuous function of r_+ as described in Figure 2(b)

The Gibbs free energy density for charged quasitopological black holes can be calculated using

$$G(T_H, \Phi) = -\int S dT_H.$$

This gives

$$\begin{aligned} G &= -\frac{1}{16\pi l^2}\left[r_+^d + \frac{d-2}{2(d-1)} l^2 \Phi^2 r_+^{d-2}\right] - \alpha \frac{d(1+d) r_+}{8\pi l^2} \\ &- \alpha \frac{(d-3)(d-2)^2}{16\pi(d-1)} \frac{\Phi^2}{r_+} + \alpha \frac{2d(d-1) r_+^2 - (d-2)^2 l^2 \Phi^2}{16\pi(d-1) l^2 r_+} \\ &\times \left(\log\left[\frac{d^2 r^{d+1}}{l^4} + \frac{(d-2)^4 \Phi^4 r_+^{d-3}}{4(d-1)^2} - \frac{d(d-2)^2 \Phi^2}{(d-1) l^2 r_+^{1-d}}\right]\right. \\ &- \left. 2\log[8\pi]\right). \end{aligned} \tag{34}$$

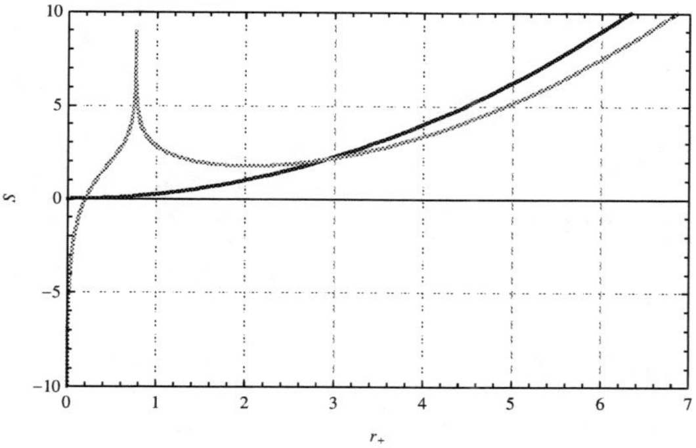

(a)

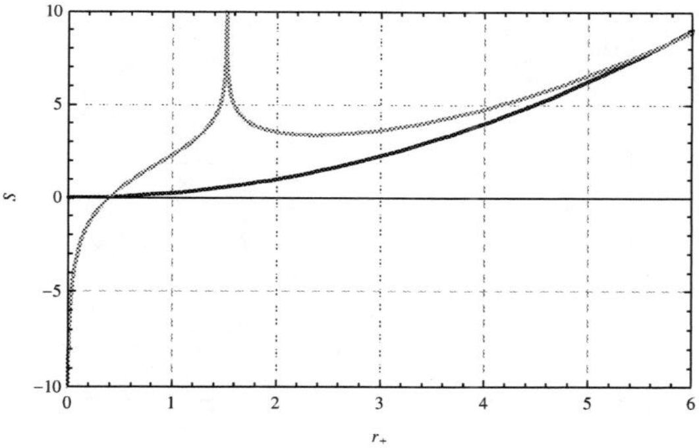

(b)

Figure 1. (a) Entropy per volume vs. the black hole horizon radius for $d = 3$, $l = 1$ and $q = 1$. (b) Entropy per volume vs. the black hole horizon radius for $d = 3$, $l = 2$ and $q = 2$. Here, $\alpha = 0$ denoted by blue line, $\alpha = 1$ denoted by green line.

In limit $\alpha \to 0$, this coincides with the original expression (16).

The effects of thermal fluctuation on the Gibbs free energy density with respect to the black hole horizon radius for $d = 3$ is depicted in figure (3). We observe that for small horizon radius the corrected entropy and original entropy of the system show opposite behavior. In fact, for a large black hole, the corrected Gibbs free energy density coincides with the original one as expected. Also, for large values of charge and AdS radius, the effect of thermal fluctuation decreases.

The charge density for the charged quasitopological black holes under the

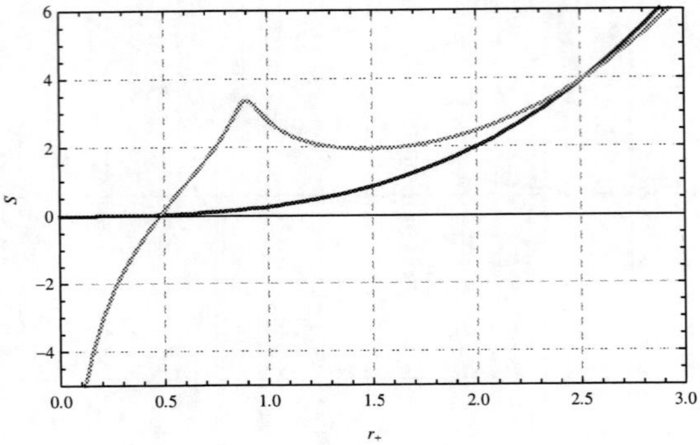

(a)

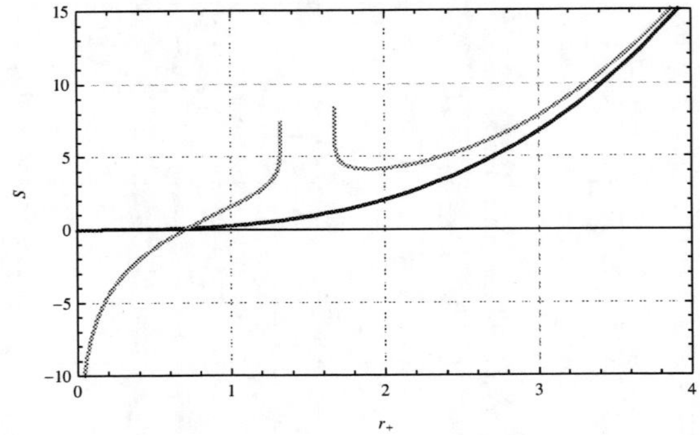

(b)

Figure 2. (a) Entropy per volume vs. the black hole horizon radius for $d = 4$, $l = 1$ and $q = 1$. (b) Entropy per volume vs. the black hole horizon radius for $d = 4$, $l = 2$ and $q = 2$. Here, $\alpha = 0$ denoted by blue line, $\alpha = 1$ denoted by green line.

influence of thermal fluctuations modifies to

$$\begin{aligned} Q &= \frac{\sqrt{2(d-1)(d-2)}}{16\pi} q \\ &- \frac{\alpha(d-2)^2}{8\pi}\sqrt{\frac{2(d-2)}{d-1}}\left[\frac{3d-(3d-1)q^2 l^2 r_+^{2-2d}}{d+(d-2)q^2 l^2 r_+^{2-2d}}\right]\frac{q}{r^{d-1}}. \end{aligned} \tag{35}$$

In limit $\alpha \to 0$, the this expression coincides with the original one (20).

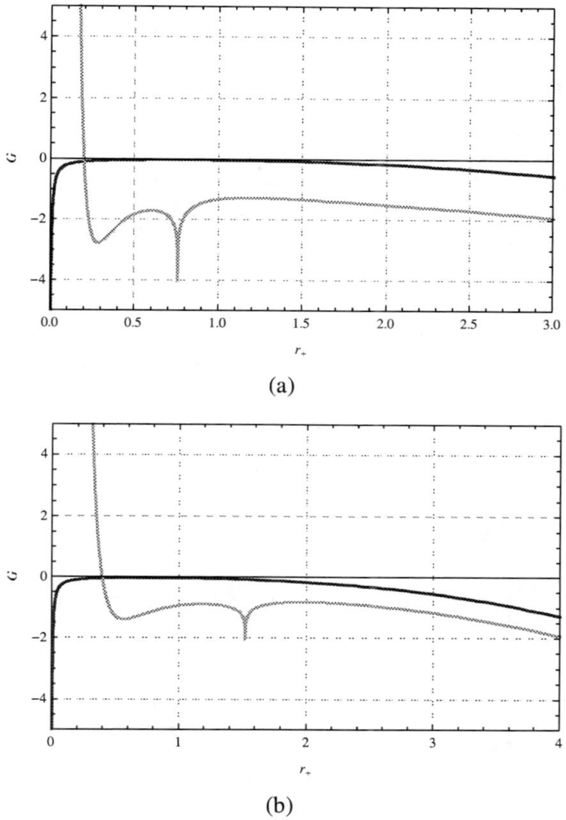

Figure 3. (a) Gibbs free energy per volume vs. the black hole horizon radius for $d = 3$, $l = 1$ and $q = 1$. (b) Gibbs free energy per volume vs. the black hole horizon radius for $d = 3$, $l = 2$ and $q = 2$. Here, $\alpha = 0$ denoted by blue line, $\alpha = 1$ denoted by green line.

With the help of above expressions, the corrected expression of the mass density for the charged quasitopological black holes is derived by

$$\begin{aligned} M &= \frac{(d-1)}{16\pi}m - \alpha\frac{d(d+1)}{8\pi l^2}r_+ \\ &- \frac{\alpha(d-2)}{8\pi}\left[\frac{d(7d-15)-(d-2)(5d-1)q^2l^2r_+^{2-2d}}{d+(d-2)q^2l^2r_+^{2-2d}}\right]\frac{q^2}{r_+^{2d-3}}. \end{aligned} \tag{36}$$

A comparative analysis of corrected and original mass densities is presented in

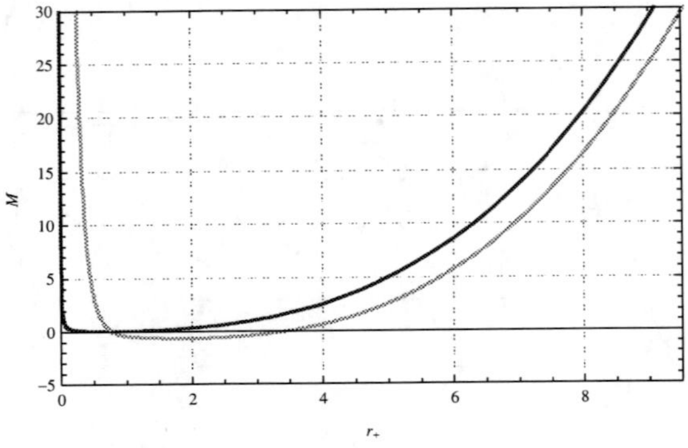

(a)

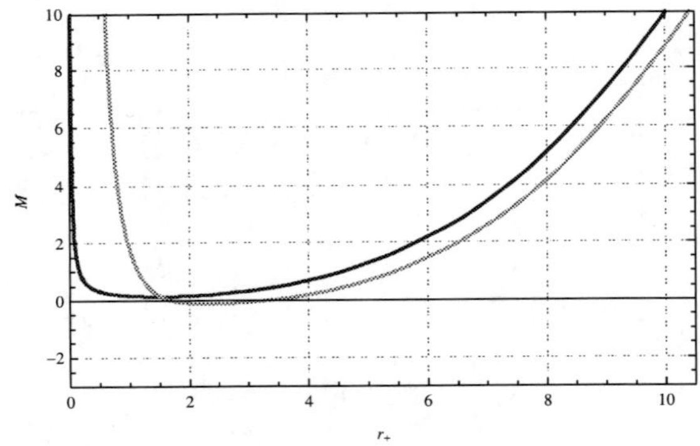

(b)

Figure 4. (a) Mass per volume vs. the black hole horizon radius for $d = 3$, $l = 1$ and $q = 1$. (b) Mass per volume vs. the black hole horizon radius for $d = 3$, $l = 2$ and $q = 2$. Here, $\alpha = 0$ denoted by blue line, $\alpha = 1$ denoted by green line.

figure (4). We can see that although there is a difference in numerical values thermal fluctuations do not change the nature of mass density even for small black holes.

4.2 Stability of Charged Quasitopological Black Holes

Now, we discuss the thermal stability of the charged quasitopological black holes under the influence of small thermal fluctuations around the equilibrium.

We, therefore, derive corrected heat capacity density under the effect of thermal fluctuation as follows:

$$
\begin{aligned}
C_\Phi &= \frac{2\pi(d-1)^2 l^2 r_+^d}{2d(d-1)r_+^2+(d-2)^2 l^2 \Phi^2} - \alpha, \\
&= \frac{\pi(d-1) l^2 r_+^{d-1}}{d r_+^{2d-2}+(d-2) l^2 q^2 r_+^{3-2d}} - \alpha. \qquad (37)
\end{aligned}
$$

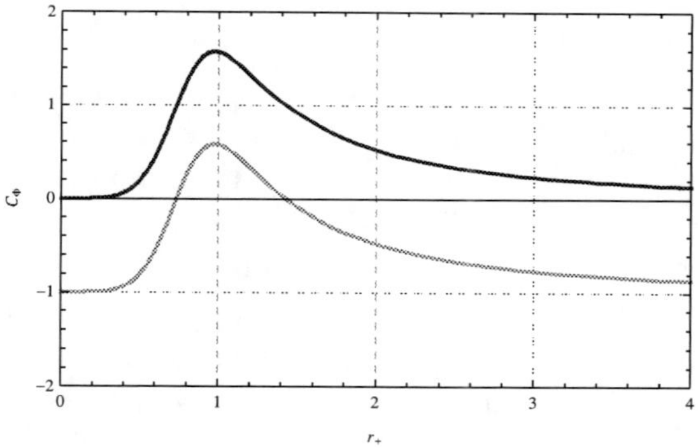

(a)

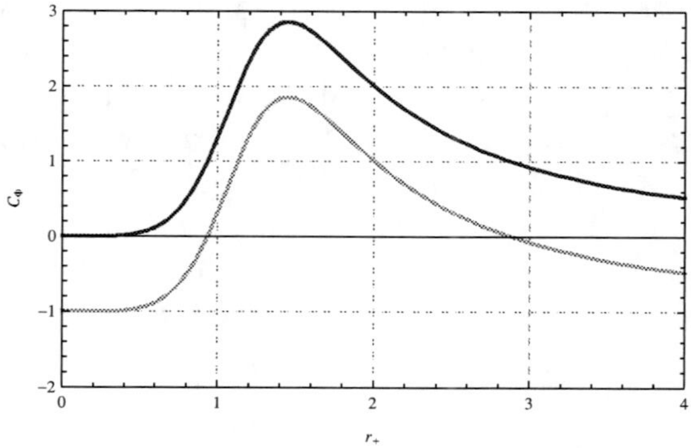

(b)

Figure 5. (a) Specific heat per volume vs. the black hole horizon radius for $d = 3$, $l = 1$ and $q = 1$. (b) Specific heat vs. the black hole horizon radius for $d = 3$, $l = 2$ and $q = 2$. Here, $\alpha = 0$ denoted by blue line, $\alpha = 1$ denoted by green line.

From the Fig. (5), we observe that for quasitopological black holes in four space-time dimensions there exists no phase transition point corresponding to both the uncorrected and corrected specific heat per volume with positive α and black holes are stable. Interestingly, we find that the correction term with negative correction parameter causes instability to the black holes and a stable state exists only in the case of corrected specific heat density with a smaller value of negative correction parameter.

Conclusion

Let us summarize the chapter. In this chapter, we have provided the basic concepts of black hole thermodynamics in general based on the Hawking's assumptions. Then, we have considered a black hole solution for the charged quasitopological gravity (which includes Gauss-Bonnet term and curvature-cubed interactions) and discussed the equilibrium thermodynamics for this. Moreover, we analyzed the stability of such black holes by computing heat capacity. Here, we found that black holes in thermal equilibrium are stable. Also, no phase transition point occurs for charged quasitopological black holes in four space-time dimensions.

Besides that we also studied the change in thermodynamics for such black holes when fluctuations around equilibrium are taken into account. Under the effect of thermal fluctuations, the microcanonical entropy is modified by an additional logarithmic term which eventually corrects all the important thermodynamical quantities. Interestingly, we observed that thermal fluctuation plays a significant role for small black holes only. In particular, we obtained a corrected version of Gibbs free energy, charge, and mass under the effects of thermal fluctuations. A modified heat capacity is also computed. To do comparative analysis, we have plotted different graphs corresponding to these quantities.

References

[1] R. M. Wald, *General Relativity*, University Press, Chicago (1984).

[2] D.Grumiller, R. McNees and J. Salzer, "Black holes and thermodynamics the first half century" in *Quantum Aspects of Black Holes*, Springer (2015).

[3] S. W. Hawking, "Black hole explosions?" *Nature* 248 30 (1974).

[4] J. D. Bekenstein, Black holes and the second law, *Lett. Nuovo Cim.* 4, 737 (1972).

[5] J. D. Bekenstein, "Black Holes and Entropy", *Phys. Rev. D* 7, 2333 (1973).

[6] S. W. Hawking, "Black holes and thermodynamics", *Phys. Rev. D* 13, 191 (1976).

[7] A. Strominger and C. Vafa, "Microscopic Origin of the Bekenstein-Hawking Entropy", *Phys. Lett. B* 379, 99 (1996).

[8] A. Ashtekar, J. Baez, A. Corichi and K. Krasnov, "Quantum Geometry and Black Hole Entropy", *Phys. Rev. Lett.* 80, 904 (1998).

[9] S. Carlip, "Black Hole Entropy from Conformal Field Theory in Any Dimension", *Phys. Rev. Lett.* 82, 2828 (1999).

[10] S. N. Solodukhin, "Conformal description of horizon's states", *Phys. Lett. B* 454, 213 (1999).

[11] S. Carlip, "Logarithmic Corrections to Black Hole Entropy from the Cardy Formula", *Class. Quant. Grav.* 17, 4175 (2000).

[12] R. B. Mann and S. N. Solodukhin, "Universality of Quantum Entropy for Extreme Black Holes", *Nucl. Phys. B* 523, 293 (1998).

[13] A. J. M. Medved and G. Kunstatter, "Quantum Corrections to the Thermodynamics of Charged 2-D Black Holes", *Phys. Rev. D* 60, 104029 (1999).

[14] A. J. M. Medved and G. Kunstatter, "One-Loop Corrected Thermodynamics of the Extremal and Non-Extremal Spinning BTZ Black Hole", *Phys. Rev. D* 63, 104005 (2001).

[15] S. N. Solodukhin, "Entropy of Schwarzschild black hole and string-black hole correspondence", *Phys. Rev. D* 57, 2410 (1998).

[16] D. Birmingham and S. Sen, "An Exact Black Hole Entropy Bound", *Phys. Rev. D* 63, 47501 (2001).

[17] S. Das, P. Majumdar and R. K. Bhaduri, "General Logarithmic Corrections to Black Hole Entropy", *Class. Quant. Grav.* 19, 2355 (2002).

[18] A. Pourdarvish, J. Sadeghi, H. Farahani, and B. Pourhassan, *Int. J. Theor. Phys.* 52, 3560 (2013).

[19] S. Upadhyay, "Quantum corrections to thermodynamics of quasitopological black holes", *Phys. Lett. B* 775, 130 (2017).

[20] S. Upadhyay, B. Pourhassan and H. Farahani, "P-V criticality of first-order entropy corrected AdS black holes in massive gravity", *Phys. Rev. D* 95, 106014 (2017).

[21] J. Jing and M. L Yan, "Statistical Entropy of a Stationary Dilaton Black Hole from Cardy Formula", *Phys. Rev. D* 63, 024003 (2001).

[22] D. Birmingham and S. Sen, "Exact black hole entropy bound in conformal field theory", *Phys. Rev. D*63, 047501 (2001).

[23] B. Pourhassan, S. Upadhyay, H. Saadat and H. Farahani, "Quantum gravity effects on Hořava-Lifshitz black hole", Nucl. Phys. B 928, 415 (2018).

[24] S. Upadhyay, "Leading-order corrections to charged rotating AdS black holes thermodynamics", Gen. Rel. Grav. 50, 128 (2018).

[25] J. M. Bardeen, B. Carter, anf S. Hawking, "The four laws of black hole mechanics", *Commun. Math. Phys.* 31, 161 (1973).

[26] D. Kubiznak, and R. B. Mann, "P-V criticality of charged AdS black holes", *JHEP* 1207, 033 (2012).

[27] R. C. Myers and B. Robinson, "Black holes in quasi-topological gravity", *J. High Energy Phys.* 08, 067 (2010).

[28] M. H. Dehghani and M. H. Vahidinia, "Surface terms of quasitopological gravity and thermodynamics of charged rotating black branes", *Phys. Rev. D* 84, 084044 (2011).

[29] M. Henningson, and K. Skenderis, "The holographic Weyl anomaly", *J. High Energy Phys.* 07 (1998) 023.

[30] M. Cvetic and S. S. Gubser, "hases of R-charged black holes, spinning branes and strongly coupled gauge theories", *J. High Energy Phys.* 04, 024 (1999).

[31] M. M. Caldarelli, G. Cognola, and D. Klemm, "Thermodynamics of Kerr-Newman-AdS Black Holes and Conformal Field Theories", *Class. Quant. Grav.* 17, 399 (2000).

[32] G. W. Gibbons and S. W. Hawking, "Action integrals and partition functions in quantum gravity", *Phys. Rev. D* 15, 2752 (1977).

[33] A. Bohr and B. R. Mottelson, *Nuclear Structure*, Vol.1 (W. A. Benjamin Inc., New York, 1969).

[34] R. K. Bhaduri, *Models of the Nucleon*, (Addison-Wesley, 1988).

Chapter 4

Motion of Spinning Particles Around Black Holes

Jose Miguel Ladino[1,*], Carlos Andrs del Valle[2,†] and Eduard Larraaga [‡1,]

[1]Universidad Nacional de Colombia, Sede Bogot
Facultad de Ciencias, Observatorio Astronmico Nacional,
Ciudad Universitaria, Bogota, Colombia

[2]Universidad Nacional de Colombia, Sede Bogot.
Facultad de Ciencias, Departamento de Fsica,
Ciudad Universitaria, Bogota, Colombia

Abstract

The motion of spinning particles around compact objects, for example a rotating stellar object moving around a supermassive black hole, is described by differential equations that are, in general, non-integrable. In this work, we present a computational code that integrates the equations of motion of a spinning particle moving in the equatorial plane of a spherically symmetric spacetime and gives a visual representation of its trajectory. This code is open source, freely available and modular, so that users may extend its application not only to the Schwarzschild metric, but also to other backgrounds.

Keywords: black holes, equations of motion, spinning particles

---

*Corresponding Author's Email: jmladinom@unal.edu.co.

†Corresponding Author's Email: cdelv@unal.edu.co.

‡Corresponding Author's Email: ealarranaga@unal.edu.co.

In: A Guide to Black Holes
Editor: Kenath Arun
ISBN: 979-8-88697-163-7

1. Introduction

Macroscopic astrophysical objects classically have a spin angular momentum associated with their rotation. When these objects, such as stars or compact objects (black holes or neutron stars), belong to physical systems (e.g. binaries), it is well known that the existence of spin has consequences on the behavior of the dynamics of the system. A particular case of interest is that of spinning bodies orbiting around supermassive black holes and its relation with the production of gravitational waves.

In order to introduce the classical spin to the description of the dynamics of extended bodies in curved spacetimes, Mathisson demonstrated in 1937 that the equations of motion show an interaction between the Riemann curvature tensor and the spin of the test particle [1]. Later, in 1951, Papapetrou made important contributions in the covariant treatment of the equations of motion, showing once again, that a spinning object moves in orbits that will differ from the geodesics [2]. Another significant input to this framework was made by Dixon in 1970, when he reformulated the equations of motion in a generalized form to hold exactly for an extended body [3]. Most of these approaches are restricted to the pole-dipole approximation, reducing the problem by considering that the body is not itself contributing to the gravitational field, and where just monopole and dipole or the mass and spin terms, respectively, are taken into account [4]. Finally, it has been studied the use of the so-called spin supplementary conditions, which fixes the center of the mass of the spinning body [5, 6, 7, 8, 9]. Following this logical path of results, the formalism that will be used in this work to describe the dynamics of a spinning particle in a curved spacetime is described by the Mathisson-Papapetrou-Dixon (MPD) equations.

Although the motion of spinning particles in black holes backgrounds is, in general, a non-integrable problem, it has been analyzed using the MPD equations under different approaches. One of the first methods considers only the equatorial motion of a spinning particle in Schwarzschild and Kerr spacetimes, and obtains the behavior of some of the properties of the innermost stable circular orbit [10]. Generalizations of this analysis, including the characteristics of black hole solutions from other gravitational theories, are presented in papers such as [11, 12, 13, 14, 15, 16, 17, 18, 19, 20, 21, 22, 23, 24, 25]. The motion of spinning particles in static and spherically symmetric backgrounds have been also studied in literature, including the Schwarzschildde Sitter spacetime [26, 27, 28], the Reissner-Nordström-de Sitter black hole [29] and in general Schwarzschild-like spacetimes [30]. Motion in backgrounds from alternative gravity theories was studied too. For example, in the background of a Schwarzschild black hole surrounded by quintessential matter field [31], in four-dimensional Einstein-Gauss-Bonnet space-time [32], around a charged Hayward black hole [33] or in a cosmological and general static spherically symmetric background [34]. All of these approaches prove that the spin of the test particle produce a non geodesic orbit, due to the existence of an additional force.

There are currently various astrophysical systems of interest that have emerged in recent years, including the collision processes between spinning particles [35, 36, 37, 20, 38, 39, 40, 41], the study of black hole backgrounds as accelerators for spinning particles [31, 21, 15, 42, 23] or the emission of gravitational radiation from spinning particles orbiting around or plunging into a black hole [43, 44, 45, 25, 14].

In this work, we provide a numerical code, which is publicly accessible and modifiable, to solve the equations of motion of a spinning particle orbiting around a spherically symmetric and static spacetime. Furthermore, it will give us the possibility of visualizing the trajectories based on the initial values of the system. This chapter is organized as follows: in Sec. 2, we present the MPD equations and the effects on the spin tensor when the motion is restricted to the equatorial plane. In Sec. 3, we introduce the metric that describe a spherically symmetric and static background to determine the components of the spin tensor. In Sec. 4, we obtain the components of the momentum vector from the associated conserved quantities of the motion. Then, the expressions of the radial and angular velocities are found in Sec. 5. Later, we show a brief summary of the step by step to solve the equations of motion computationally in Sec. 6. and finally, in Sec. 7, the code is commented.

2. Equations of Motion for a Spinning Particle

2.1. The Mathisson-Papapetrou-Dixon Equations

The MPD equations describe the motion of a spinning particle in a curved background. They state that [1, 2, 3]

$$\begin{cases} \frac{DP^\mu}{D\lambda} = & -\frac{1}{2}R^\mu{}_{\nu\alpha\beta}u^\nu S^{\alpha\beta} \\ \frac{DS^{\mu\nu}}{D\lambda} = & P^\mu u^\nu - u^\mu P^\nu, \end{cases} \tag{1}$$

where P^μ and $u^\mu = \frac{dx^\mu}{d\lambda}$ are the momentum and velocity vectors of the test particle, $R^\mu{}_{\nu\alpha\beta}$ is the Riemann tensor and $S^{\mu\nu}$ represents the spin tensor that defines the spin angular momentum of the test particle through the relation

$$s^2 = \frac{1}{2}S^{\mu\nu}S_{\mu\nu}. \tag{2}$$

As is well known, the proper mass of the test particle is given by the norm of the momentum vector,

$$m^2 = -P_\alpha P^\alpha. \tag{3}$$

but, introducing the quantity

$$\mu = -P^\beta u_\beta, \tag{4}$$

equations (1) gives the relation

$$P^\alpha = \mu u^\alpha - u^\beta \frac{DS^{\alpha\beta}}{D\lambda}, \tag{5}$$

indicating that the momentum and the velocity vectors are no longer parallel for spinning particles. This fact gives the freedom to introduce an additional condition which may produce different trajectories [12, 5]. There are many equations that we can introduce, but we will specifically use the *Tulczyjew spin-supplementary condition* [46, 5], which restricts the spin tensor to generate only rotations through the relation

$$P_\mu S^{\mu\nu} = 0. \tag{6}$$

Due to the spherical symmetry of the background, it is possible to consider, without loosing generality, that the orbital motion is restricted to the equatorial plane, $\theta = \frac{\pi}{2}$, which implies $P_2 = 0$. For simplicity, we will only consider the spin aligned or anti-aligned orbits. Thus, we impose the condition $S^{2\mu} = 0$ in equation (6), to obtain the non-vanishing components of the spin tensor as

$$\begin{cases} S^{01} = -S^{10} = & \frac{P_3}{P_0} S^{13} \\ S^{03} = -S^{30} = & -\frac{P_1}{P_0} S^{13}. \end{cases} \tag{7}$$

3. Spherically Symmetric Static Spacetimes

In this work, we will consider a spherically symmetric and static spacetime, with the following general line element

$$ds^2 = g_{00}dt^2 + g_{11}dr^2 + g_{22}d\theta^2 + g_{33}d\phi^2. \tag{8}$$

Using this metric, together with equations (6) and (7), the spin angular momentum of the test particle is written as

$$s^2 = \left[g_{00}g_{11}P_3^2 + g_{00}g_{33}P_1^2 + g_{11}g_{33}P_0^2\right] \frac{(S^{13})^2}{P_0^2}. \tag{9}$$

The determinant of the metric tensor is given by

$$g = \det g_{\mu\nu} = g_{00}g_{11}g_{22}g_{33} \tag{10}$$

while the inverse tensor is calculated as

$$g^{\mu\nu} = \frac{1}{g} \begin{pmatrix} g_{11}g_{22}g_{33} & 0 & 0 & 0 \\ 0 & g_{00}g_{22}g_{33} & 0 & 0 \\ 0 & 0 & g_{00}g_{11}g_{33} & 0 \\ 0 & 0 & 0 & g_{00}g_{11}g_{22} \end{pmatrix}. \tag{11}$$

Then, using these results we obtain

$$s^2 = -\, m^2(g_{00}g_{11}g_{33})\frac{(S^{13})^2}{P_0^2}. \tag{12}$$

Solving for the spin tensor component S^{13} and using this result in equations (7) gives

$$\begin{cases} S^{13} = & -\frac{sP_0}{m} \frac{1}{\sqrt{-g_{00}g_{11}g_{33}}} \\ S^{01} = & -\frac{sP_3}{m} \frac{1}{\sqrt{-g_{00}g_{11}g_{33}}} \\ S^{03} = & -\frac{sP_1}{m} \frac{1}{\sqrt{-g_{00}g_{11}g_{33}}}. \end{cases} \tag{13}$$

4. Conserved Quantities in the Motion of a Spinning Test Particle

Given the existence of a Killing vector k, we associate the existence of a conserved quantity in the motion of the spinning particle through the relation

$$C_k = P^\mu k_\mu + \frac{1}{2} S^{\mu\nu} \nabla_\mu k_\nu. \tag{14}$$

Assuming that the metric tensor components do not depend explicitly on the coordinates t and ϕ, there are two Killing vectors,

$$\xi = \partial_t \qquad \text{and} \qquad \zeta = \partial_\phi, \tag{15}$$

corresponding to the conservation of energy and angular momentum, respectively. For the first of these vectors we have

$$C_t = -E = \xi^\mu P_\mu + \frac{1}{2} S^{\mu\nu} \nabla_\mu \xi_\nu. \tag{16}$$

The covariant derivative in the second term gives

$$S^{\mu\nu} \nabla_\mu \xi_\nu = S^{\mu\nu} \partial_\mu \xi_\nu - S^{\mu\nu} \Gamma^\alpha_{\mu\nu} \xi_\alpha = S^{\mu\nu} \partial_\mu \xi_\nu \tag{17}$$

and because $\xi_\mu = g_{\mu\nu} \xi^\nu = g_{00} \delta^0_\mu$, we obtain

$$E = -P_0 - \frac{1}{2} S^{\mu 0} \partial_\mu g_{00}. \tag{18}$$

Considering the dependence $g_{00} = g_{00}(r)$, we get

$$E = -P_0 - \frac{1}{2} S^{10} \partial_r g_{00} \tag{19}$$

and using equations (13) this becomes

$$E = -P_0 - \frac{1}{2} \frac{sP_3}{m} \frac{\partial_r g_{00}}{\gamma}, \tag{20}$$

where we introduce $\gamma = \sqrt{-g_{00} g_{11} g_{33}}$. A similar procedure for the Killing vector ζ gives

$$C_\phi = J = \zeta^\mu P_\mu + \frac{1}{2} S^{\mu\nu} \nabla_\mu \zeta_\nu = \zeta^\mu P_\mu + \frac{1}{2} S^{\mu\nu} \partial_\mu \zeta_\nu \tag{21}$$

This time, the Killing vector satisfies $\zeta_\mu = g_{\mu\nu}\zeta^\nu = g_{33}\delta^3_\mu$, and therefore

$$J = P_3 + \frac{1}{2}S^{\mu 3}\partial_\mu g_{33}. \tag{22}$$

Considering now the dependence $g_{33} = g_{33}(r)$ (because considering that the motion occurs in the equatorial plane implies that there is no dependence with the coordinate θ), we obtain

$$J = P_3 + \frac{1}{2}S^{13}\partial_r g_{33} \tag{23}$$

and using equations (13) this becomes

$$J = P_3 - \frac{1}{2}\frac{sP_0}{m}\frac{\partial_r g_{33}}{\gamma}. \tag{24}$$

From equations (20) and (24) we solve to obtain the components P_0 and P_3 of the momentum vector as

$$\begin{cases} P_0 = & -\frac{2m\gamma(2m\gamma E + sJ\partial_r g_{00})}{4m^2\gamma^2 + s^2\partial_r g_{00}\partial_r g_{33}} \\ P_3 = & \frac{2m\gamma(2m\gamma J + sE\partial_r g_{33})}{4m^2\gamma^2 + s^2\partial_r g_{00}\partial_r g_{33}}. \end{cases} \tag{25}$$

The component P_1 is obtained from the normalization condition (3), giving

$$(P_1)^2 = -\frac{m^2 + g^{00}(P_0)^2 + g^{33}(P_3)^2}{g^{11}}. \tag{26}$$

5. The Equations of Motion Revisited

Now, we are able to write explicitly the equations of motion given in (1). Using a parameter λ for which $\dot{u}^0 = 1$, we have that the equation for the component S^{01} of the spin tensor is

$$\frac{DS^{01}}{D\lambda} = P^0 u^1 - P^1. \tag{27}$$

Using equations (7) and (13) in the left hand side gives

$$-\frac{sP_3}{m}\frac{D}{D\lambda}\left(\frac{1}{\gamma}\right) - \frac{s}{m\gamma}\frac{DP_3}{D\lambda} = P^0\dot{r} - P^1. \tag{28}$$

Since the factor $\frac{1}{\gamma} = \frac{1}{\sqrt{-g_{00}g_{11}g_{33}}}$ depends only on the coordinate r, we have that

$$\frac{D}{D\lambda}\left(\frac{1}{\gamma}\right) = \frac{dx^\mu}{d\lambda}\nabla_\mu\left(\frac{1}{\gamma}\right) = \frac{dr}{d\lambda}\partial_r\left(\frac{1}{\gamma}\right) = -\frac{\dot{r}}{\gamma^2}\partial_r\gamma. \tag{29}$$

Replacing in the equation of motion gives

$$\dot{r}\left[P^0 - \frac{sP_3}{m\gamma^2}\partial_r\gamma\right] = P^1 - \frac{s}{m\gamma}\frac{DP_3}{D\lambda}. \tag{30}$$

The last term in the right hand side can be evaluated using the equation of motion for P_3 arising from (1),

$$\frac{DP_3}{D\lambda} = -\frac{1}{2}R_{3\nu\alpha\beta}u^\nu S^{\alpha\beta}. \tag{31}$$

This gives the final result

$$\dot{r}\left[P^0 - \frac{sP_3}{m\gamma^2}\partial_r\gamma\right] = P^1 + \frac{s}{2m\gamma}R_{3\nu\alpha\beta}u^\nu S^{\alpha\beta}. \tag{32}$$

A similar procedure using the equation of motion for the component S^{03} of the spin tensor, gives the following relation,

$$\dot{\phi}P^0 + \frac{sP_1}{m\gamma^2}\dot{r}\partial_r\gamma = P^3 - \frac{s}{2m\gamma}R_{1\nu\alpha\beta}u^\nu S^{\alpha\beta}. \tag{33}$$

Note that using the anti-symmetry properties (in the first two indices) of the Riemann tensor and the fact that the particle is moving on the equatorial plane $\dot{\theta} = 0$, we have that

$$R_{3\nu\alpha\beta}u^\nu S^{\alpha\beta} = R_{30\alpha\beta}S^{\alpha\beta} + R_{31\alpha\beta}\dot{r}S^{\alpha\beta} \tag{34}$$

and

$$R_{1\nu\alpha\beta}u^\nu S^{\alpha\beta} = R_{10\alpha\beta}S^{\alpha\beta} + R_{13\alpha\beta}\dot{\phi}S^{\alpha\beta}. \tag{35}$$

Therefore, the equations of motion are now

$$\begin{cases} \dot{r}\left[P^0 - \frac{sP_3}{m\gamma^2}\partial_r\gamma\right] = & P^1 + \frac{s}{2m\gamma}R_{30\alpha\beta}S^{\alpha\beta} + \frac{s}{2m\gamma}R_{31\alpha\beta}S^{\alpha\beta}\dot{r} \\ \dot{\phi}P^0 + \frac{sP_1}{m\gamma^2}\dot{r}\partial_r\gamma = & P^3 - \frac{s}{2m\gamma}R_{10\alpha\beta}S^{\alpha\beta} - \frac{s}{2m\gamma}R_{13\alpha\beta}S^{\alpha\beta}\dot{\phi}. \end{cases} \tag{36}$$

From the first of these relations, we obtain an expression for the radial velocity, $\dot{r}$,

$$\dot{r} = \frac{P^1 + \frac{s}{2m\gamma} R_{30\alpha\beta} S^{\alpha\beta}}{P^0 - \frac{sP_3}{m\gamma^2}\partial_r\gamma - \frac{s}{2m\gamma} R_{31\alpha\beta} S^{\alpha\beta}}, \tag{37}$$

while the second relation gives an equation for the angular velocity $\dot{\phi}$,

$$\dot{\phi} = \frac{P^3 - \frac{sP_1}{m\gamma^2}\dot{r}\partial_r\gamma - \frac{s}{2m\gamma} R_{10\alpha\beta} S^{\alpha\beta}}{P^0 + \frac{s}{2m\gamma} R_{13\alpha\beta} S^{\alpha\beta}}. \tag{38}$$

6. Summary of the Equations and Procedure to Solve Them Computationally

In order to solve the equations of motion for the particular case of the spinning particle in the equatorial plane of a spherically symmetric and static spacetime, we will impose initial values for the following quantities:

- Proper mass of the particle: m (conserved)
- Initial position: t_0, r_0 and ϕ_0
- Total Energy of the particle: E (conserved)
- Orbital Angular Momentum : j (conserved)
- Spin Angular Momentum : s (conserved)

The total angular momentum is determined by the values of j and s by

$$J = j + s. \tag{39}$$

Given these initial values, we proceed with the following steps:

1. Obtain the initial values for the components of the momentum using the equations

$$\begin{cases} P_0 = & -\frac{2m\gamma(2m\gamma E + sJ\partial_r g_{00})}{4m^2\gamma^2 + s^2\partial_r g_{00}\partial_r g_{33}} \\ P_3 = & \frac{2m\gamma(2m\gamma J + sE\partial_r g_{33})}{4m^2\gamma^2 + s^2\partial_r g_{00}\partial_r g_{33}} \\ (P_1)^2 = & -\frac{m^2 + g^{00}(P_0)^2 + g^{33}(P_3)^2}{g^{11}}. \end{cases} \tag{40}$$

2. Determine the initial values of the spin tensor components

$$\begin{cases} S^{13} = & -\frac{sP_0}{m\gamma} \\ S^{01} = & -\frac{sP_3}{m\gamma} \\ S^{03} = & -\frac{sP_1}{m\gamma} \end{cases} . \tag{41}$$

3. Calculate the Riemann tensor components $R_{\mu\nu\alpha\beta}$.

4. Numerically solve the equations

$$\dot{r} = \frac{P^1 + \frac{s}{2m\gamma} R_{30\alpha\beta} S^{\alpha\beta}}{P^0 - \frac{sP_3}{m\gamma^2} \partial_r \gamma - \frac{s}{2m\gamma} R_{31\alpha\beta} S^{\alpha\beta}} \tag{42}$$

and

$$\dot{\phi} = \frac{P^3 - \frac{sP_1}{m\gamma^2} \dot{r} \partial_r \gamma - \frac{s}{2m\gamma} R_{10\alpha\beta} S^{\alpha\beta}}{P^0 + \frac{s}{2m\gamma} R_{13\alpha\beta} S^{\alpha\beta}} \tag{43}$$

to obtain the new values of the coordinates r and ϕ.

5. Begin the process again at step 1.

7. The MSPBH Code

The MSPBH (Motion of Spinning Particles around Black Holes) Code[1] is a Python implementation that solves the equations of motion described above to obtain the trajectory of a spinning particle around a spherically symmetric compact object described by a metric tensor. The program uses symbolic algebra packages SymPy and EinsteinPy to calculate the required quantities, such as partial derivatives of the metric and the Riemann tensor, and the integration of the equations of motion is performed using odeint from the SciPy package. The code reads all the necessary variables, constants, initial conditions, and information about the metric needed to calculate the trajectories from a configuration file. When executing the code without configuration, a file called template.txt is created with a usage example corresponding to the last stable orbit in the Schwarzschild metric.

[1]The MSPBH Code is available at https://github.com/cdelv/MSPBH

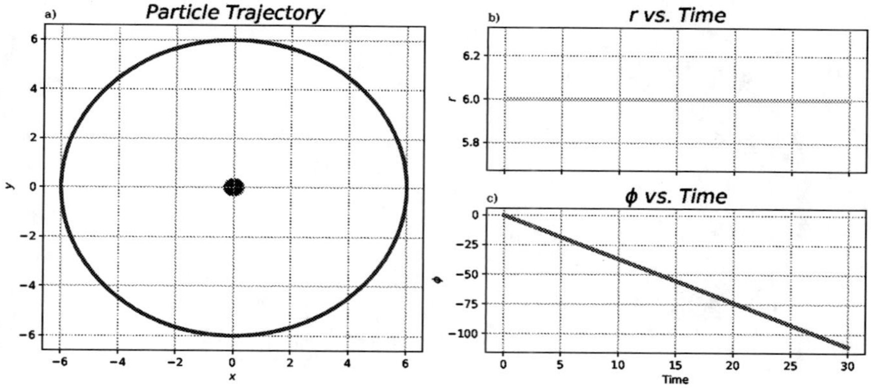

Figure 1. Trajectory of a particle with mass $m = 1$, around a black hole with mass $M = 1$. The initial conditions are $r_0 = 6$, $\phi_0 = 0$, the particle energy is $E = -\sqrt{8/9}$, the angular momentum is $j = \sqrt{12}$ and the spin is $s = 0$.

Some of the parameters and options in the configuration file include the definition of the metric itself. For example, if the options User_tensor and User_metric are set to FALSE, the program will use the incorporated Schwarzschild metric from the library Einsteinpy, and compute the Riemann tensor. When User_tensor is TRUE, the program will use the components of the tensors given by the user on the configuration file (this option has precedence over User_metric. In fact, when User_metric is TRUE and User_tensor is FALSE, the program will use the function Create_User_Metric_Tensor to create a custom metric tensor with the library Einsteinpy. If User_metric is False, you don't have to give the Riemann and metric tensor components on the template. In this case, once the metric tensor is determined, the code will compute the Riemann tensor using the EinsteinPy library and proceeds to calculate the orbit.

A successful execution of the code will provide a series of plots (is the user sets options Plot_trajectoryy and/or Plot_energy to TRUE), including the trajectory of the particle and the behavior of the coordinates $r(t)$ and $\phi(t)$, and two output files: 'log.out' and 'data.csv'. The first one has information about all the variables read from the configuration file, while the second is a CSV file where you will find the information about time, r and ϕ coordinates, energy, and angular momentum of the particle at each

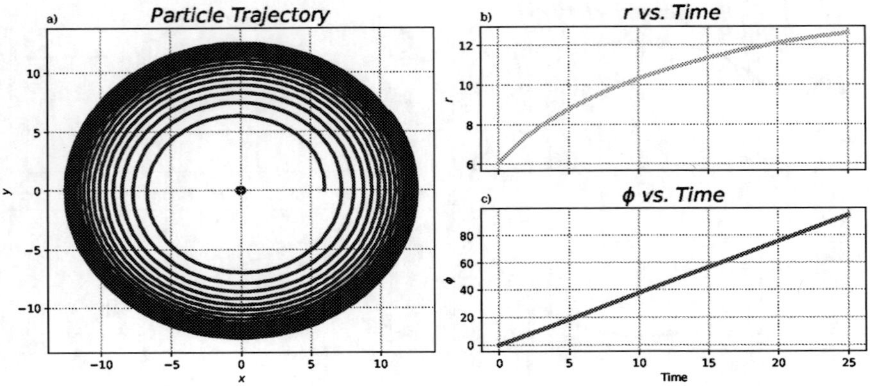

Figure 2. Trajectory of a particle with mass $m = 1$, around a black hole with mass $M = 1$. The initial conditions are $r_0 = 6$, $\phi_0 = 0$, the particle energy is $E = -\sqrt{8/9}$, the angular momentum is $j = \sqrt{12}$ and the spin is $s = 3$.

step in the integration. In Figure 1, the trajectory and the evolution of the coordinates of a non spinning particle can be seen.

For a spinning particle there are two possible configurations: the spin aligns with the angular momentum of the orbit or not. In Figure 2 we show the trajectory of the same particle described in Figure 1 but this time with a non-zero spin aligned with the angular momentum.

Conclusion

We have considered the equations of motion of a spinning particle moving in the equatorial plane of a spherically symmetric and static background to write the expressions for the radial and angular velocities in terms of the components of the momentum vector, the spin tensor and the Riemann tensor. After given a brief summary of the step by step algorithm to solve these equations of motion numerically, we have presented a Python implementation named the MSPBH Code, which solves numerically the equations of motion and provides a visualization of the trajectory. The code also gives the coordinates of the spinning particle together with the corresponding energy and angular momentum of the spinning particle.

We have illustrated the implementation by considering two examples (non spinning particle and spinning particle moving around a Schwarzschild black hole) and showing the obtained trajectory. The results confirm that the spin of the particle causes a change in the trajectory, producing a fall into the black hole.

The MSPBH Code provides a practical and visual tool to demonstrate how the spin of the particle affects its own dynamics in a spherically symmetric and static spacetime. It is open source, freely available and modular, so that users may extend its application not only to the default schwarzschild metric, but also to other backgrounds included in the EinsteinPy library or even by defining its own metric.

Acknowledgments

This work was supported by the Universidad Nacional de Colombia. Research Incubator No. 64-Computational Astrophysics and Hermes Grant Code 41673.

References

[1] Mathisson M., Das zitternde elektron und seine dynamik. *Acta Phys. Pol*, 6:218, 1937.

[2] Papapetrou A., Spinning Test-Particles in General Relativity. I. *Proceedings of the Royal Society of London Series A*, 209(1097):248–258, October 1951.

[3] Dixon W. G., Dynamics of Extended Bodies in General Relativity. I. Momentum and Angular Momentum. *Proceedings of the Royal Society of London Series A*, 314(1519):499–527, January 1970.

[4] Toshmatov Bobir and Malafarina Daniele, Spinning test particles in the γ spacetime. *Phys. Rev. D*, 100:104052, Nov 2019.

[5] Lukes-Gerakopoulos Georgios, Seyrich Jonathan, and Kunst Daniela, Investigating spinning test particles: spin supplementary conditions and the Hamiltonian formalism. *Phys. Rev. D*, 90(10):104019, 2014.

[6] Timogiannis Iason, Lukes-Gerakopoulos Georgios, and Apostolatos Theocharis A., Spinning test body orbiting around a schwarzschild black hole: Comparing spin supplementary conditions for circular equatorial orbits. *Physical Review D*, 104(2), Jul 2021.

[7] Lukes-Gerakopoulos Georgios, Time parameterizations and spin supplementary conditions of the Mathisson-Papapetrou-Dixon equations. *Phys. Rev. D*, 96:104023, Nov 2017.

[8] Mikóczi Balázs, Spin supplementary conditions for spinning compact binaries. *Phys. Rev. D*, 95:064023, Mar 2017.

[9] Filipe L., Costa O. and Natário José, *Center of Mass, Spin Supplementary Conditions, and the Momentum of Spinning Particles*, pages 215–258. Springer International Publishing, Cham, 2015.

[10] Suzuki Shingo and Maeda Kei-ichi, Innermost stable circular orbit of a spinning particle in kerr spacetime. *Physical Review D*, 58(2), Jun 1998.

[11] Tanaka Takahiro, Mino Yasushi, Sasaki Misao, and Shibata Masaru, Gravitational waves from a spinning particle in circular orbits around a rotating black hole. *Phys. Rev. D*, 54:3762–3777, Sep 1996.

[12] Zhang Yu-Peng, Wei Shao-Wen, Guo Wen-Di, Sui Tao-Tao, and Liu Yu-Xiao, Innermost stable circular orbit of spinning particle in charged spinning black hole background. *Phys. Rev. D*, 97:084056, Apr 2018.

[13] Zhang Ming and Liu Wen-Biao, Innermost stable circular orbits of charged spinning test particles. *Physics Letters B*, 789:393–398, 2019.

[14] Lukes-Gerakopoulos Georgios, Harms Enno, Bernuzzi Sebastiano, and Nagar Alessandro, Spinning test body orbiting around a kerr black hole: Circular dynamics and gravitational-wave fluxes. *Physical Review D*, 96(6), Sep 2017.

[15] Zhang Yu-Peng, Gu Bao-Min, Wei Shao-Wen, Yang Jie, and Liu Yu-Xiao, Charged spinning black holes as accelerators of spinning particles. *Phys. Rev. D*, 94:124017, Dec 2016.

[16] Pugliese Daniela, Quevedo Hernando, and Ruffini Remo, Equatorial circular orbits of neutral test particles in the kerr-newman spacetime. *Phys. Rev. D*, 88:024042, Jul 2013.

[17] Conde Carlos, Galvis Cristian, and Larrañaga Eduard, Properties of the innermost stable circular orbit of a spinning particle moving in a rotating maxwell-dilaton black hole background. *Phys. Rev. D*, 99:104059, May 2019.

[18] Jefremov Paul I., Tsupko Oleg Yu., and Bisnovatyi-Kogan Gennady S., Innermost stable circular orbits of spinning test particles in schwarzschild and kerr space-times. *Phys. Rev. D*, 91:124030, Jun 2015.

[19] Nucamendi Ulises, Becerril Ricardo, and Sheoran Pankaj, Bounds on spinning particles in their innermost stable circular orbits around rotating braneworld black hole. *The European Physical Journal C*, 80(1):35, Jan 2020.

[20] Liu Yunlong and Zhang Xiangdong, Maximal efficiency of the collisional penrose process with spinning particles in kerr-sen black hole. *The European Physical Journal C*, 80(1):31, Jan 2020.

[21] An Jincheng, Peng Jun, Liu Yan, and Feng Xing-Hui, Kerr-sen black hole as accelerator for spinning particles. *Phys. Rev. D*, 97:024003, Jan 2018.

[22] Guo Minyong and Gao Sijie, Kerr black holes as accelerators of spinning test particles. *Physical Review D*, 93(8), Apr 2016.

[23] Zhang Songming, Liu Yunlong, and Zhang Xiangdong, Kerr–de sitter and kerr–anti–de sitter black holes as accelerators for spinning particles. *Phys. Rev. D*, 99:064022, Mar 2019.

[24] Kubizňák David and Cariglia Marco, Integrability of spinning particle motion in higher-dimensional rotating black hole spacetimes. *Phys. Rev. Lett.*, 108:051104, Jan 2012.

[25] Han Wen-Biao, Gravitational radiation from a spinning compact object around a supermassive kerr black hole in circular orbit. *Phys. Rev. D*, 82:084013, Oct 2010.

[26] Mortazavimanesh M. and Mohseni Morteza, Spinning particles in schwarzschild–de sitter space–time. *General Relativity and Gravitation*, 41(11):2697–2706, Nov 2009.

[27] Kunst Daniela, Perlick Volker, and Lämmerzahl Claus, Isofrequency pairing of spinning particles in schwarzschild–de sitter spacetime. *Phys. Rev. D*, 92:024029, Jul 2015.

[28] Plyatsko Roman, Panat Volodymyr, and Fenyk Mykola, Nonequatorial circular orbits of spinning particles in the schwarzschild–de sitter background. *General Relativity and Gravitation*, 50(11):150, Oct 2018.

[29] Stuchlík Zden ěk and Hledík Stanislav, Equilibrium of a charged spinning test particle in reissner-nordström backgrounds with a nonzero cosmological constant. *Phys. Rev. D*, 64:104016, Oct 2001.

[30] Mohseni Morteza, Stability of circular orbits of spinning particles in schwarzschild-like space–times. *General Relativity and Gravitation*, 42(10):2477–2490, Oct 2010.

[31] Sheoran Pankaj, Nandan Hemwati, Hackmann Eva, Nucamendi Ulises, and Abebe Amare, Schwarzschild black hole surrounded by quintessential matter field as an accelerator for spinning particles. *Phys. Rev. D*, 102:064046, Sep 2020.

[32] Zhang Yu-Peng, Wei Shao-Wen, and Liu Yu-Xiao, Spinning test particle in four-dimensional einsteingaussbonnet black holes. *Universe*, 6(8), 2020.

[33] Larraaga Eduard, Circular motion and the innermost stable circular orbit for spinning particles around a charged hayward black hole background. *International Journal of Modern Physics D*, 29(16):2050121, 2020.

[34] Zalaquett Nicolas, Hojman Sergio A., and Asenjo Felipe A., Spinning massive test particles in cosmological and general static spherically symmetric spacetimes. *Classical and Quantum Gravity*, 31(8):085011, apr 2014.

[35] Maeda Kei-ichi, Okabayashi Kazumasa, and Okawa Hirotada, Maximal efficiency of the collisional penrose process with spinning particles. *Phys. Rev. D*, 98:064027, Sep 2018.

[36] Mukherjee Sajal, Collisional penrose process with spinning particles. *Physics Letters B*, 778:54–59, 2018.

[37] Zhang Ming, Jiang Jie, Liu Yan, and Liu Wen-Biao, Collisional penrose process of charged spinning particles. *Phys. Rev. D*, 98:044006, Aug 2018.

[38] Yuan Xulong, Liu Yunlong, and Zhang Xiangdong, Collision of spinning particles near BTZ black holes. *Chinese Physics C*, 44(6):065104, jun 2020.

[39] Hackmann Eva, Nandan Hemwati, and Sheoran Pankaj, Particle collisions near static spherically symmetric black holes. *Physics Letters B*, 810:135850, 2020.

[40] Liu Yan and Liu Wen-Biao, Energy extraction of a spinning particle via the super penrose process from an extremal kerr black hole. *Phys. Rev. D*, 97:064024, Mar 2018.

[41] Armaza Cristóbal, Banados Máximo, and Koch Benjamin, Collisions of spinning massive particles in a schwarzschild background. *Classical and Quantum Gravity*, 33(10):105014, apr 2016.

[42] Zaslavskii O. B., Schwarzschild black hole as particle accelerator of spinning particles. *EPL (Europhysics Letters)*, 114(3):30003, may 2016.

[43] Saijo Motoyuki, Maeda Kei-ichi, Shibata Masaru, and Mino Yasushi, Gravitational waves from a spinning particle plunging into a kerr black hole. *Phys. Rev. D*, 58:064005, Aug 1998.

[44] Mino Yasushi, Shibata Masaru, and Tanaka Takahiro, Gravitational waves induced by a spinning particle falling into a rotating black hole. *Phys. Rev. D*, 53:622–634, Jan 1996.

[45] Harms Enno, Lukes-Gerakopoulos Georgios, Bernuzzi Sebastiano, and Nagar Alessandro, Asymptotic gravitational wave fluxes from a

spinning particle in circular equatorial orbits around a rotating black hole. *Phys. Rev. D*, 93:044015, Feb 2016.

[46] Hojman Sergio A. and Asenjo Felipe A., Can gravitation accelerate neutrinos? *Classical and Quantum Gravity*, 30(2):025008, dec 2012.

Chapter 5

Testing a Microscopic Model for Black Holes Deduced from Maximum Force

Christoph Schiller*
Motion Mountain Research, Munich, Germany

Abstract

Maximum force $c^4/4G$ and maximum speed c imply Einstein's field equations of general relativity. Combining the two limits with the quantum of action and an old idea by Dirac leads to a Planck-scale model for the microscopic degrees of freedom of space, particles and horizons. The model is based on fluctuating, tangled, one-dimensional strands. The model reproduces black hole mass, energy, temperature and entropy. Strands confirm that the mass and charge of a black hole are distributed over its horizon. The model implies black hole radiation, the hoop conjecture, the laws of black hole mechanics, the no-hair theorem, Bekenstein's entropy bound, the lack of remnants, as well as both the magnetic moment limit and the g-factor of black holes. Comparisons with fuzzballs and firewalls are possible. Important predictions are numbered; so far, all agree with observations or expectations. In addition, the standard model of particle physics arises automatically from fluctuating strands.

Keywords: general relativity, black holes, quantum gravity, strand conjecture

*Corresponding Author's Email: cs@motionmountain.net

In: A Guide to Black Holes
Editor: Kenath Arun
ISBN: 979-8-88697-163-7

1. Introduction: The Search for the Microscopic Degrees of Freedom of Black Holes

What are the microscopic degrees of freedom of black holes? Whatever they are, they must be related to the microscopic degrees of freedom of vacuum, and to gravitons. Over the past decades, many candidates have been proposed [1, 2, 3, 4, 5, 6, 7, 8, 9, 10], though with limited success. Any proposed microscopic degrees of freedom of black holes must *reproduce* space, curvature, mass and gravitation in all their macroscopic and microscopic aspects. But in addition, any proposal must also provide *new* results that go beyond the usual description of space as a continuous manifold made of points. In the following, it will be shown how an idea on quantum theory due to Dirac, combined with a recent result on general relativity, leads to a microscopic model of black holes that provides such new results. The other quantum gravity aspects of the model have been explored in detail elsewhere [11, 12].

2. The First Origin of the Strand Conjecture: Tethers

Bohr had the habit to present quantum theory as a consequence of the smallest action value $\hbar$ [13]. In 1925, Dirac included the speed limit c into quantum theory. From 1929 on, Dirac presented the *string trick* or *belt trick* in his lectures. The belt trick is illustrated in Figure 1. It describes the basic properties of spin $1/2$ – namely the return to the original situation ofter a rotation by 4π – as the consequence of *tethering*. As Dirac told Gardner ([14], page 47) the belt trick also shows that a spin value below $\hbar/2$ is impossible. This confirms Bohr's statement: a smallest angular momentum value $\hbar/2$ also limits the smallest observable action to the value $\hbar$.

A less known companion to the belt trick, the fermion trick, is illustrated in Figure 2. The fermion trick describes the basic properties of fermions – the return to the original situation ofter double particle exchange – as the consequence of tethering. Here, *tethers* are connections of a structure to spatial infinity with unobservable, uncuttable, flexible, fluctuating, one-dimensional strands.

In nature, matter particles are found to have spin $1/2$ and to behave as fermions. However, particle tethers are not observed at all. In retrospect, Dirac's tether trick was the first hint that matter particles could be described using *unobservable* tethers with *observable* crossing switches. (The concept of 'cross-

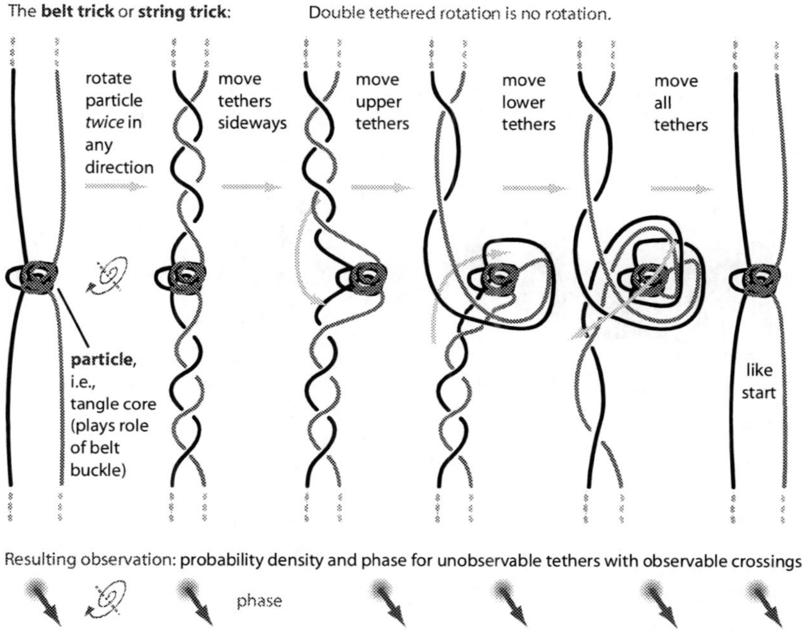

Figure 1. The *belt trick* or *string trick*: a double rotation, by 4π, of a tethered structure, such as a tangle core, is equivalent to no rotation – if the tethers are allowed to fluctuate and untangle as shown. Untangling is impossible after a particle rotation by just 2π: tangle cores with 4 or more tethers thus show the properties of spin $1/2$. As a consequence, a tethered core is able to rotate continuously. In the strand conjecture, the trick also couples rotation and *displacement* of the tangle core. The coupling of core rotation frequency and core displacement determines particle mass. (The figure is modified from reference [15].)

ing switch' is defined below.) Fifty years later, in 1980, Battey-Pratt and Racey showed that tethers also imply the full free Dirac equation [16]. In short, Battey-Pratt and Racey proved that Dirac's trick implies Dirac's equation. (For example, strand crossings have the same geometrical properties as wave functions: position, density, and phase. Wave functions turn out to be time-averages of tether crossings [11, 15].) In other terms, *every quantum effect* can be seen as the result of observable crossing switches of unobservable tethers.

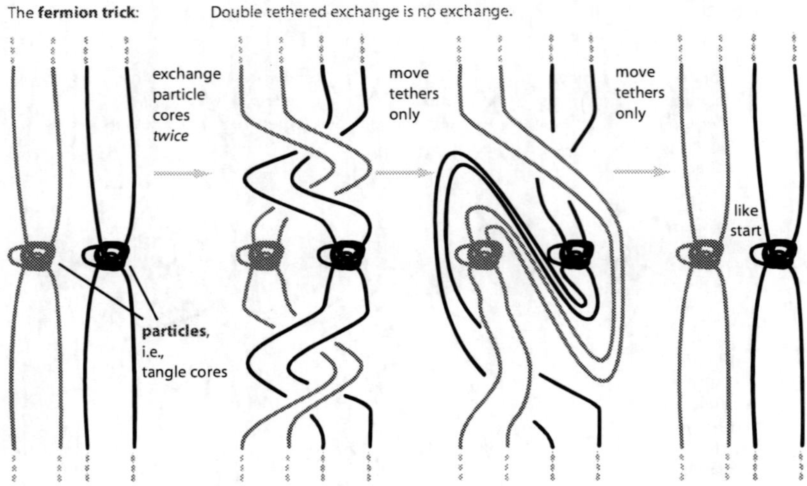

Figure 2. The *fermion trick*: a double particle exchange of two tethered particles is equivalent to no exchange – if the tethers are allowed to fluctuate and untangle as shown. Untangling is impossible after single exchange. Tangle cores with 4 or more tethers thus show the defining properties of fermions.

3. The Second Origin of the Strand Conjecture: Maximum Force

The invariant limit speed c realized by massless radiation is a fundamental principle of nature. Indeed, special relativity contains and implies a maximum speed; furthermore, the maximum speed implies and contains special relativity. In a similar way, the invariant maximum force

$$F_{\mathrm{max}} = \frac{c^4}{4G} \approx 3.0 \cdot 10^{43}\,\mathrm{N} \ , \qquad (1)$$

that is realized by gravitational horizons is a principle of nature. Here, G is the gravitational constant. The force limit is a principle of nature for two analogous reasons.

It is known since at least 1973 that *general relativity contains and implies a maximum force* [17, 18, 19, 20, 21, 22, 23, 24, 25, 26, 27, 28, 29, 30, 31, 32, 33, 34, 35, 36, 37]. The gravitational force between two black holes never exceeds $c^4/4G$, as shown by Gibbons [23]. The force limit is also given by the maximum energy per distance possible in general relativity: the energy Mc^2 of a Schwarzschild black hole divided by its diameter $D = 4GM/c^2$ defines the maximum possible force value. The ratio is independent of the black hole size and mass. Even charged or rotating black holes do not yield larger ratios. Finally, the force on a test mass being lowered towards a gravitational horizon with a rope never exceeds the upper bound, – if the size of the test mass is considered. In fact, no physical system allows exceeding the upper force limit. All proposed counter-examples to the force limit disappear under closer scrutiny [38, 39, 40, 41, 42]. In fact, the force bound agrees with all known observations and passes all theoretical tests [42].

It is known since 2003 that the force bound $c^4/4G$ *implies and contains* Einstein's field equations. Two proofs are known [24, 25, 43, 42]. One proof starts by deducing the first law of horizon mechanics from the force bound. The first law can be used to deduce the field equations. The other proof starts by showing that the force bound implies a limit on the deformation of space. The deformation limit in turn implies a relation between curvature and energy, which leads to the field equations.

In short, because the field equations *follow* from maximum force $c^4/4G$ and from maximum speed c, both limits are *principles* of nature. It should be mentioned that alternative and *equivalent* formulations of the principle of maximum force exist. There is the principle of maximum power or maximum luminosity $c^5/4G$, the principle of maximum mass flow rate $c^3/4G$, and others [42]. Each of these principles is equivalent to general relativity, and equivalent to the others.

4. Combining Tethers and Maximum Force

Taken together, the three limits – special relativity's c, quantum theory's $\hbar$, and general relativity's $c^4/4G$ – imply that there are no trans-Planckian effects in nature of any kind (as long as G is substituted by $4G$ in all Planck quantities). These (corrected) Planck limits and all their consequences agree with all observations. Therefore, the (corrected) Planck limits must also hold in a theory of quantum gravity.

In fact, there is a simple way to ensure the lack of trans-Planckian effects. One requires that the smallest quantum of action $\hbar$, the smallest length $\sqrt{4G\hbar/c^3}$ and the smallest time $\sqrt{4G\hbar/c^5}$ hold everywhere and at every instant of time.

Can unobservable strands also explain gravitation and quantum gravity? It turns out that this is the case, provided that the fluctuating strands have Planck radius. With this property, the validity of the maximum force principle is ensured, and thus the validity of general relativity. Also quantum gravity follows, including the finite value and the surface dependence of black hole entropy, as shown below. It thus appears that also *every gravitational effect* can be seen as the result of observable crossing switches of unobservable tethers.

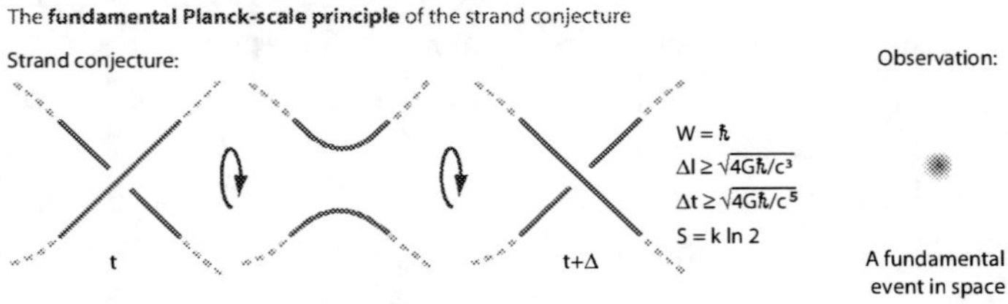

Figure 3. The fundamental principle of the strand conjecture specifies the simplest observation – a simplified version of Dirac's trick, shown in Figure 1, taking place at the Planck scale – that is possible in nature: the almost point-like *fundamental event* results from a *skew strand switch*, or *crossing switch*, at a position in three-dimensional space. The strands themselves are *not observable*. They are impenetrable and are best imagined as having Planck size radius. The observable switch defines the action unit $\hbar$. The double Planck length limit and the double Planck time limit arise, respectively, from the smallest and from the fastest crossing switch possible. The paper plane represents background space, i.e., the local tangent Euclidean space defined by the observer. (The figure is modified from reference [15].)

5. The Fundamental Principle of the Strand Conjecture

In the strand conjecture, all physical systems found in nature – matter, radiation, space and horizons – are made of unobservable strands that fluctuate at the Planck scale.

▷ A *strand* is defined as smooth curved line – a one-dimensional, open, continuous, everywhere infinitely differentiable subset of $\mathbb{R}^3$ or of a curved 3-dimensional Riemannian space, with trivial topology and without endpoints – that is surrounded by a perpendicular disk of Planck radius $\sqrt{\hbar G/c^3}$ at each point of the line, whose shape is not self-intersecting, and that is randomly fluctuating over time.

(It is most practical to define and visualize strands as having Planck-size radius. A few issues about this choice are not discussed here.) The definition of strands leads to the following statements:

- ▷ Strands are unobservable. However, crossing switches of skew strands – exchanges of over- and underpasses – are observable. Crossing switches determine the Planck units G, c and $\hbar$; this fundamental principle is illustrated in Figure 3.

The defining Figure 3 therefore combines the essence of Dirac's trick with the Planck limits, and thus combines quantum theory with maximum force and general relativity.

The strand conjecture claims that the fundamental principle contains and implies all observations, all equations of motion, and all Lagrangians. In particular, the fundamental principle implies:

- ▷ Physical space is a (three-dimensional) *network* of fluctuating strands – i.e., of strands that are neither woven nor tangled nor knotted, as illustrated in Figure 4).
- ▷ Horizons are (two-dimensional) *weaves* of fluctuating strands – i.e., similar to a fabric made of woven threads, and illustrated in Figure 6.
- ▷ Particles are (localized) *rational tangles* of fluctuating strands – using the term from topological knot theory, defined and illustrated in Figure 10.
- ▷ Physical motion *minimizes* the number of observable crossing switches of fluctuating unobservable strands.

The fundamental principle thus contains all physical systems, and also every physical process. This includes particle motion, gauge interactions, particle scattering, and gravitational waves. Classifying tangles of strands leads to the particle spectrum, classifying tangle deformations with Reidemeister moves leads to the three gauge groups and couplings, and exploring tether exchanges leads to particle mixings. The belt trick leads to particle mass. The full Lagrangian of the standard model follows from strands, including massive neutrino with PMNS mixing [11, 15]. In the following, however, only gravitation is explored, and in particular, only black holes.

By definition, the fundamental principle of the strand conjecture of Figure 3 states that action, length, time and entropy are *limited from below*:

$$W \geq \hbar \ , \quad \Delta l \geq \sqrt{4G\hbar/c^3} \ , \quad \Delta t \geq \sqrt{4G\hbar/c^5} \ , \quad S \geq k \ln 2 \ . \qquad (2)$$

Above all, strands visualize these inequalities. Together with Figure 3, these inequalities contain quantum gravity and allow understanding black holes. The number $\ln 2$ in the minimum entropy is due to the 2 possible strand configurations.

Strands themselves have *no* observable properties: no colour, no tension, no mass, and no energy. Due to the impossibility of observing them, strands have *no* meaningful equation of motion. Indeed, all results in the following are *independent* of the detailed fluctuating motion one might imagine for strands. (This could be called the *congruence principle* of the strand conjecture.) The independence of fluctuations eliminates any imaginable arbitrariness of the description of space, horizons and particles with strands.

Strands *cannot be cut*; they are not made of parts. Strands cannot interpenetrate; they *never* form an actual crossing. When the term ‘crossing’ is used in the present context, only the two-dimensional projection shows a crossing. Strands thus have a *crossing* in space when a strand segment passes over another. In three dimensions, strands are *always at a distance*. Like in Dirac’s trick, a *crossing switch* – the change from an overpass to an underpass – cannot arise through strand interpenetration, but only via strand deformation.

In the strand conjecture, *all physical observables* – such as energy, mass, action, momentum, length, velocity, surface, volume, entropy, field intensities or quantum numbers – arise from combinations of crossing switches. No physical observable is a property of strands themselves; all physical observables arise from *shape configurations* of *several* strands. In other terms: every physical observable *emerges* from strand crossings.

6. Flat and Curved Physical Space from Strands

The *only* observable aspect of strands – as in Dirac’s trick – are their crossing switches, and thus, for example, the distribution of crossing switches. To relate strands to observations, it is important to deduce the behaviour of crossing switches from the fundamental principle for the system under consideration. The simplest case is empty physical space.

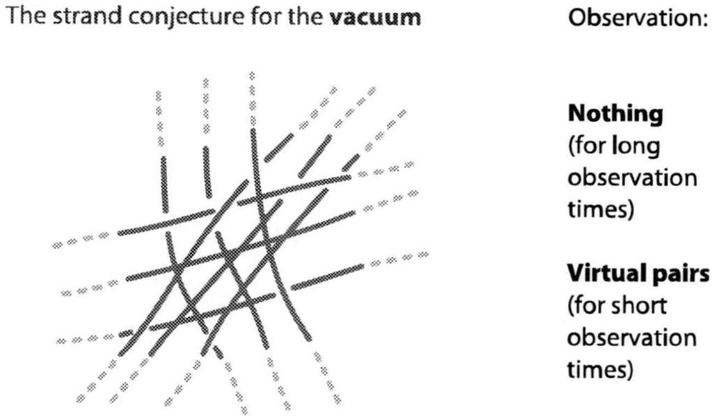

Figure 4. A simplified and idealized illustration of the strand conjecture for a flat vacuum, i.e., for flat physical space. The space of the picture is *background* space. *Physical* space is generated by strand crossing and their switches. Strands fluctuate in all directions: their crossings are washed out. (Typical strand distances can be many orders of magnitude larger than their diameters.) For sufficiently long time scales (longer than a few Planck times), the lack of crossing switches leads to a vanishing energy density; for short time scales, particle–antiparticle pairs, i.e., rational tangle–antitangle pairs, arise in the vacuum due to the shape fluctuations of the strands. (The figure is modified from reference [12]; the reference also discusses the relation between background space and physical space.)

A strand network of untangled strands describing physical space is illustrated in Figure 4. The illustration uses background space to define physical space. *Background* space is what is needed to *talk* about nature. *Physical* space is what can be *measured* about space: curvature, vacuum energy, entropy, temperature etc. The circularity issues that arise are discussed and solved in reference [12].

In other terms, a network of *untangled*, *unwoven* and *unknotted* strands represents *empty and flat* physical space. The time-average of the fluctuations, taken on a scale of a few Planck times or more, yields three-dimensional physical flat space, including its continuity, homogeneity, isotropy and Lorentz-invariance. On time scales longer than a few Planck times, there are (on average) *no* crossing switches, and thus neither matter nor energy – just empty space. *Physical space results from washing out the crossing switches of the*

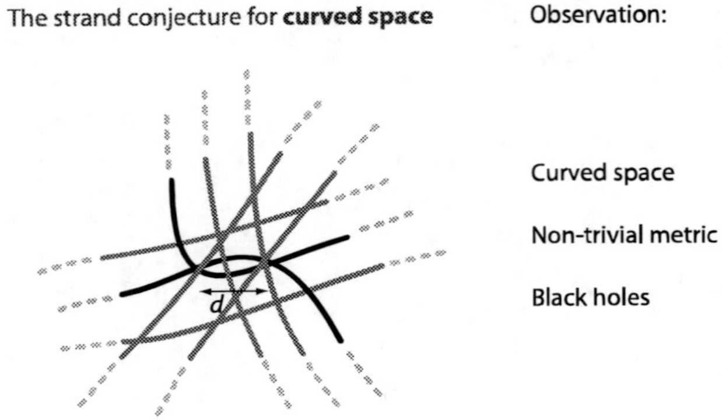

Figure 5. An illustration of the strand conjecture for a *curved* vacuum. The strand and crossing configuration is *not* homogeneous and is midway between that of a flat vacuum and that of a horizon. Strands in black differ in their configuration from those in a flat vacuum. The value of the curvature is inversely proportional to the distance d. (The figure is modified from reference [12].)

strand network. Strands thus imply that *no deviation* from the continuity, homogeneity, isotropy, dimensionality and Lorentz-invariance of (physical) flat space can be observed. This is the case at any practical energy (except for the corrected Planck energy itself) and is valid *despite* the existence of a smallest length $l_{\min} = \sqrt{4G\hbar/c^3}$.

Strands not only visualize flat space; strands also visualize curved space. Spatial curvature is illustrated in Figure 5. In simple terms, the fundamental principle of the strand conjecture implies:

- ▷ *Flat space* is a homogeneous network of crossing fluctuating strands.
- ▷ *Curvature* is an inhomogeneous crossing (switch) density in the vacuum network.

The strand configuration for curved space differs from that of flat space: certain strands break the isotropy and homogeneity. The main curvature value depends on the configuration of the strands leading to the inhomogeneity. The curvature can evolve over time. This strand model for curved space implies that curved space-time is, locally, a Minkowski space. Thus, strands lead to a *pseudo-Riemannian space-time*.

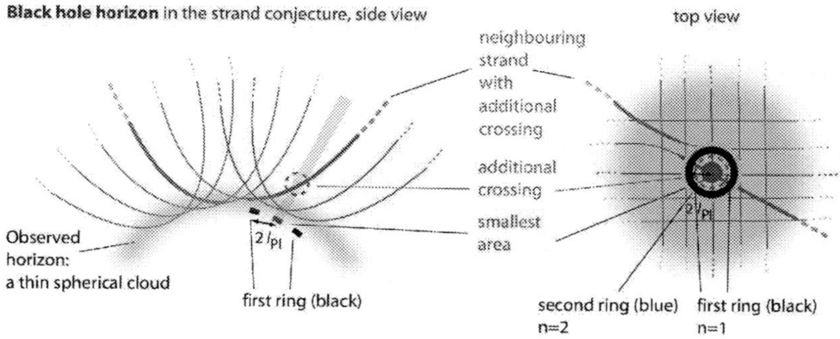

Figure 6. The strand conjecture is illustrated for a Schwarzschild black hole, as seen by a distant observer: the horizon is a cloudy or fuzzy surface produced by crossing switches of the strands woven *tightly* into it. Due to the additional crossings *above* the horizon, the number of microstates per smallest area is larger than 2, and given by the base e of the natural logarithms (see text). This horizon model yields the entropy of black holes. (The figure is modified from reference [12].)

7. Horizons and Black Holes from Strands

In the strand conjecture,

▷ Horizons are one-sided, tight *weaves* of strands.

In this statement, *one-sided* means that all strands leave the horizon on one side, the side of the observer. One-sidedness means that there is 'nothing', not even an unobservable strand, on the other side of the horizon – except for rare fluctuations. For a distant observer *at rest*, a one-sided weave also implies that no space and no events are observable behind it. The weave thus acts as a *limit* to observation. Figure 6 gives a schematic illustration of a Schwarzschild black hole, both as a cross section and as a top view. For any black hole horizon, all strands come in from far away, are *woven* into the horizon, and leave again to far away. Due to their Planck radius, the weave of strands forming a horizon is as *tight* as possible: seen from above the horizon, there is one crossing for each smallest area.

For a *falling* observer, the strands do not form a weave, but continue on the other side and form a (distorted) network, i.e., curved vacuum. Such an observer does not notice anything special when approaching the horizon, or when crossing it. A tangle weave thus shows all the qualitative properties that characterize a horizon.

The weave model for horizons allows determining the *energy* and thus the *mass* of a Schwarzschild black hole. Energy E has the dimension action per time. Because every crossing switch is associated with an action $\hbar$, the horizon energy is found by determining the number $N_{\rm cs}$ of crossing switches, multiplied by $\hbar$, that occur per unit time. Because the number will depend on the surface area of the horizon, the energy will be proportional to horizon area. In a horizon weave, the crossing switches *propagate* from one crossing to the next, over the surface of the whole (tight) weave. Because the horizon weave is *tight*, each crossings has the size of the minimum length squared, given by the corrected Planck area $A_{\rm cPl} = 4\,G\hbar/c^3$. Because the horizon weave is *tight*, the propagation speed is one smallest length per shortest switch time: switch propagation thus occurs at the speed of light c. In the time T needed to circumnavigate a *spherical*, non-rotating horizon of area $A = 4\pi R^2$ at the speed of light, *all* crossings of the horizon switch. This yields:

$$E = \frac{N_{\rm cs}\hbar}{T} = \frac{A/(4G/c^3)}{2\pi R/c} = \frac{c^4}{2G}\,R\ . \tag{3}$$

The weave model of a horizon thus reproduces the relation between the energy – or mass – and the radius of a Schwarzschild black hole.

The weave model also fixes the number of microstates per horizon area. Figure 6 shows that for a given smallest area containing just one strand crossing, the effective number N of microstates *above* that smallest area is *larger* than 2. A number larger than 2 occurs because sometimes, fluctuating neighbouring strands also *cross above* the smallest area.

The probability for a neighbouring strand to cross above the smallest area will depend on the distance at which the neighbouring strand leaves the lowest woven layer of the horizon. To calculate the probability, one imagines the central crossing surrounded by an infinite series of rings, each with a smallest area value $A_{\rm cPl} = 4\,G\hbar/c^3$. The rings can be numbered with a number n, as illustrated in Figure 6. The central crossing corresponds to $n = 0$. Ring number n thus *encloses* an area given by the smallest area $A_{\rm cPl}$ times n. Now, the probability p_1 that a strand from ring 1 crosses above the centre is

$$p_1 = \frac{1}{2} = \frac{1}{2!}\ . \tag{4}$$

Likewise, the probability that a strand from ring n crosses above the centre is

$$p_n = \frac{1}{n+1}\, p_{n-1} = \frac{1}{(n+1)!} \ , \tag{5}$$

because the strand has to continue in the correct direction above every ring on its way to the centre. The last expression is a result of the *extension* of strands; it would not arise if the fundamental constituents of horizons were not extended. The expression yields an effective number N of microstates above the central crossing given by

$$N = 2 + \frac{1}{2!} + \frac{1}{3!} + \frac{1}{4!} + ... + \frac{1}{n!} + ... = \mathrm{e} = 2.718281... \tag{6}$$

In this expression, the term 2 is due to the two options at the central point; the term $1/2!$ arises from the first ring around it, as illustrated in Figure 6; the subsequent terms are due to the subsequent rings. Expression (6) implies that the average number N of strand microstates for each smallest area, i.e., for each *corrected Planck area* $A_{\mathrm{cPl}} = 4\,G\hbar/c^3$ on the black hole horizon, is given by $N = \mathrm{e} > 2$. In the weave model, every corrected Planck area therefore contains somewhat *more* than 1 bit of information.

The calculation of the entropy of the complete black hole horizon is straightforward. It starts with the definition of entropy

$$S = k\ \ln N_{\mathrm{total}} \ , \tag{7}$$

where k is the Boltzmann constant and N_{total} is total number of microstates of the complete horizon. The full horizon area A can be seen as composed of *a large number* of corrected Planck areas. The total number of microstates is thus the product of the number N of microstates for every corrected Planck area:

$$N_{\mathrm{total}} = N^{A/(4\,G\hbar/c^3)} \ . \tag{8}$$

This is standard thermodynamics. The next step is to insert the result (6) due to strands. This yields

$$N_{\mathrm{total}} = \mathrm{e}^{A/(4\,G\hbar/c^3)} \ . \tag{9}$$

The total number of horizon microstates, inserted into expression (7), yields the horizon entropy S of a black hole with surface A:

$$\frac{S}{k} = \frac{A}{4\,G\hbar/c^3} \ . \tag{10}$$

This is the expression discovered by Bekenstein [44].

To summarize: in the strand conjecture, black hole entropy is *finite* because the microscopic degrees of freedom are discrete. Black hole entropy is due to one quarter of surface area A because the microscopic degrees of freedom are extended. The factor $1/4$ is due to the same factor appearing in the maximum force value $c^4/4G$. As Figure 6 illustrates, strands imply that horizon entropy is located *at and slightly around* the horizon. This agrees with expectations. The strand conjecture for black holes also confirms and visualizes a result by Zurek and Thorne from 1985: the entropy of a black hole is the logarithm of the number of ways in which it could have been made [45].

In the strand conjecture, the above calculation of the black hole entropy counts certain states more than once. Because strands can bend, reorienting the complete horizon sphere does not produce a different micro-state. The possible orientations of a sphere are given by the possible orientations of the poles and by the possible orientations of the sphere around the pole axis. The poles of the sphere can point to any of the A/A_{cPl} minimal surfaces that make up the horizon; in addition, the sphere can be rotated around the axis in $\sqrt{A/A_{\mathrm{cPl}}} \cdot \mathcal{O}(1)$ ways. The corrected value for the number of microstates of a spherical horizon is therefore

$$N_{\mathrm{total}} = \frac{N^{A/A_{\mathrm{cPl}}}}{(A/A_{\mathrm{cPl}})^{3/2} \cdot \mathcal{O}(1)} \quad . \tag{11}$$

This value yields the corrected black hole entropy

$$\frac{S}{k} = \frac{A}{4\,G\hbar/c^3} - \frac{3}{2}\ln\frac{A\,c^3}{4G\hbar} - \ln\mathcal{O}(1) \quad . \tag{12}$$

The last term is negligibly small. The second term shows that the strand conjecture makes a specific prediction for the logarithmic correction to the entropy of a Schwarzschild black hole. The value of the correction is much too small to ever be tested in experiments; but it agrees with previous calculations [46].

In other terms, woven strands imply both the energy E and the entropy S of Schwarzschild black holes. As usual, the ratio $E/2S$ determines the black hole *temperature* [47]:

$$T_{\mathrm{BH}} = \frac{\hbar c}{4\pi k}\,\frac{1}{R} = \frac{\hbar}{2\pi k c}\,a \quad . \tag{13}$$

In the last equality, the surface gravitational acceleration $a = GM/R^2 = c^2/2R$ was introduced, using expression (3).

In short, black holes are also warm. Their finite temperature implies that *black holes radiate.* Strands thus reproduce black hole *evaporation.* Radiation and evaporation are due to strands detaching from the horizon. If a single strand detaches, a photon is emitted. If a tangle of two or three strands detaches, a graviton or a fermion is emitted. When all strands have detached, the full black hole has evaporated.

8. Deducing General Relativity from Thermodynamics and Strands

In a path-breaking paper, Jacobson showed in 1995 that the thermodynamic properties of the microscopic degrees of freedom of space and of black holes imply Einstein's field equations of general relativity [48]. He deduced the result without requiring any details about the nature and properties of these degrees of freedom.

Jacobson started with three thermodynamic properties:

- the entropy–area relation $S = A\,kc^3/4G\hbar$,
- the temperature–acceleration relation $T = a\,\hbar/2\pi kc$,
- the relation between heat and entropy $\delta Q = T\delta S$.

As shown above, these three relations are valid also for strands: strands reproduce the entropy relation (7) of black holes, the temperature (13) of black holes, and their heat–entropy relation from (3).

Using the three thermodynamic properties, the thermodynamic relation

$$\delta E = \delta Q \ , \tag{14}$$

which is valid *only* in case of a horizon, yields

$$\delta E = \frac{c^2}{8\pi G} a\,\delta A \ . \tag{15}$$

This is the first law of black hole mechanics. This law can be rewritten, using the energy–momentum tensor T_{ab}, as

$$\int T_{ab}\,k^a \mathrm{d}\Sigma^b = \frac{c^2}{8\pi G} a\,\delta A \ , \tag{16}$$

where $d\Sigma^b$ is the general surface element and k is the Killing vector that generates the horizon. The Landau–Raychaudhuri equation [49] – a purely geometric relation – allows rewriting the right-hand side as

$$\int T_{ab}\, k^a d\Sigma^b = \frac{c^4}{8\pi G} \int R_{ab}\, k^a d\Sigma^b \ , \tag{17}$$

where R_{ab} is the Ricci tensor that describes space-time curvature. The equality between integrals implies a relation between the integrands:

$$T_{ab} = \frac{c^4}{8\pi G} \left(R_{ab} - (\frac{R}{2} + \Lambda)\, g_{ab} \right) \ , \tag{18}$$

where R is the Ricci scalar and Λ is a constant of integration. These are Einstein's field equations of general relativity, without any modification. The thermodynamics of horizons thus do not determine the value of the cosmological constant Λ, as expected.

Jacobson stresses that the field equations are valid everywhere and for all times – not only near horizons – because a suitable general coordinate transformation can position a horizon at any point in space-time. If horizons and black holes are thermodynamic systems, so is curved space itself. It is thus correct to say that the field equations result from *thermodynamics of space*. Jacobson's argument thus shows that space, gravity and horizons are made of the same microscopic degrees of freedom.

Given that strands realize the premises of Jacobson's argument, it is possible to formulate a first, basic prediction:

1. *No deviations* between general relativity and the strand conjecture arise.

This prediction also implies the full validity of the Hilbert Lagrangian, without any modification. As a result, all processes described by general relativity are reproduced by strands. For example, this includes universal $1/r^2$ gravity; it is illustrated in Figure 7.

So far, the *cosmological horizon* and its effects were not considered. Strands thus imply that general relativity holds without any measurable deviation and with full certainty only for *sub-galactic* distances, where the horizon has *no* influence. The validity of the strand model for galactic and cosmological distances will be explored in the future.

The equivalence of strands and general relativity is not a surprise: strands and their properties were deduced from maximum force, which already implies general relativity. In fact, the equivalence is not a strong statement for a further reason. Jacobson's deduction of the field equations is *independent* of the microscopic model of space. After Jacobson's result, various kinds of microscopic degrees of freedom for space have been conjectured and explored [1, 2, 3, 4, 5, 6, 7, 8, 9, 10]. All these proposals recover the field equations. In other terms, these explorations have shown that extracting the *correct* microscopic degrees of freedom of space from all the proposals in the literature is *not possible* using arguments from quantum gravity alone.

The correct microscopic degrees of freedom of space and gravitation must also reproduce the standard model of particle physics. In addition, they must explain particle masses coupling constants and mixing angles. These seem *the only criteria* that differentiate between the various microscopic models of gravitation. Interestingly, strands are one of the few approaches to quantum gravity that do reproduce the Lagrangian of the standard model, as shown in detail elsewhere [11, 15].

9. Strand Predictions about Physical Space

2. In the network that defines empty space, the tangling of strands is not possible in other dimensions or in fundamentally different ways. Therefore, strands predict that flat physical space is *three-dimensional*, *unique* and *well-behaved* at all scales. Flat physical space is a three-dimensional *continuum* that is *homogenous* and *isotropic*, without observable deviations. Curved space is *Riemannian*. These predictions agree both with expectations [50] and with the most recent observations [51, 52].

   Any evidence for lower or higher dimensions, other spatial topologies, quantum foam, different vacuum states, domain walls, cosmic strings, regions of negative energy, 'space-time noise', 'particle diffusion', 'space viscosity', crystal behaviour of space, or any other deviation from a well-behaved pseudo-Riemannian space-time manifold would directly *falsify* the strand conjecture.

3. As a consequence of the fundamental principle the maximum local energy speed in nature is c. Strands predict this at all energy scales, in all directions, at all times, at all positions, for every physical observer. Strand

predict the lack of *observable* violations of Lorentz-invariance, for all energies and all physical systems. It predicts no variable speed of light, no time-dependent speed of light, no time-dependent energy of light, i.e., no 'tired' light, no energy-dependent speed of light, no helicity-dependent speed of light, no polarization-dependent speed of light, no 'double relativity' and no 'deformed special relativity'. Strands thus predict the lack of dispersion, birefringence and opacity of the vacuum. All this agrees with observations [53].

4. Strand crossings resemble fermionic or anti-commuting coordinates as used in supergravity, resemble non-commutative space [54, 55], resemble Clifford algebras, and even resemble the internal spaces of the *aikyon* approach based on octonions [56]. A strand crossing can also be seen as a four-dimensional subspace, spanned by the four angles describing the crossing geometry, specific to a point in background space; a crossing thus resembles twistor space [57]. Though strands resemble all these types of internal spaces, they do so only at *certain* points in space and at *certain* instants in time, as a result of the fluctuations. Figure 4 and Figure 5thus imply that strands do *not* produce fixed internal spaces.

5. The strand conjecture for the vacuum predicts the *lack of trans-Planckian effects*. If *any effect* due to space intervals smaller than the minimal length can be observed – for example in electric dipole moments [58], in higher order effects in quantum field theory, or in the discreteness of space – the strand conjecture is falsified. The same holds for time intervals shorter than the corrected Planck time.

   As a consequence, strands also imply a *limit* to curvature κ:

$$\kappa \leq \frac{1}{l_{\mathrm{min}}} = \sqrt{\frac{c^3}{4G\hbar}} \; . \tag{19}$$

   In particular, this limit implies the *lack of singularities* in nature, of any kind. Physical space is Riemannian in all observations.

10. Strand Predictions about Black Holes

For black holes, the fundamental principle of the strand conjecture leads to numerous consequences.

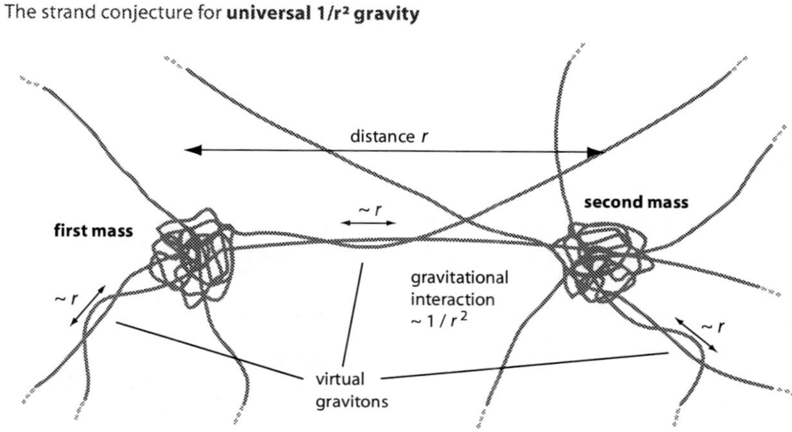

Figure 7. Gravitational attraction results from strands. More precisely, everyday *gravitation* is due to tether pair twists and their influence on tether fluctuations. When speeds are low and spatial curvature is negligible, as illustrated here, twisted tether pairs – i.e., virtual gravitons – from any mass lead to a universal $1/r^2$ attraction of other masses. The average length of twisted pairs of tethers scales with r. As a consequence, the curvature around such a mass scales as $1/r^3$. These results are valid for infinite, approximately flat space. The strand structure of gravitons is illustrated in Figure 9. (The figure is taken from reference [12].)

6. No black hole merger can exceed maximum power $c^5/4G$. Indeed, the highest luminosity values observed so far are those observed in black hole mergers by LIGO and VIRGO [31]. At present, the highest peak powers were observed for the events GW170729 and GW190521. They showed values of $4.2(1.5) \cdot 10^{49}$ W or 230 ± 80 solar masses per second [59] and of $3.7(9) \cdot 10^{49}$ W or 207 ± 50 solar masses per second [60]. All these values are well below the (corrected) Planck limit of $c^5/4G = 50\,756(12)$ solar masses per second.

7. The strand conjecture implies that a *rotating* black hole realizes a belt trick that involves a *vast* number of tethers. Figure 8 shows such a configuration during rotation. Animations illustrating the process were available on the internet before the strand conjecture formulated this equivalence,

programmed by Jason Hise. In the figure, the *ergosphere* is the spatial region in which the crossing switches take place during rotation. The figure does not show that the horizon of a rotating black hole is somewhat flattened at the poles.

Figure 8. The strand conjecture leads to an unusual model of a black hole rotating about the vertical axis (© Jason Hise). The flattening at the poles is not shown. For a complete animation, see the online videos at youtube.com/watch?v=LLw3BaliDUQ and youtube.com/watch?v=eR9ZCwYPhhU. The figure also confirms the moment of inertia of Schwarzschild black holes and the g-factor of rotating charged black holes (see text).

8. The strand conjecture for black holes illustrated in Figure 6 implies that the horizon entropy, the horizon energy and the horizon temperature are *limit values* for all physical systems of the same size. These limits arise directly from the corrected Planck limits of expressions (2) that define the strand conjecture. So far, they agree with observations.

   In particular, because strands cannot be tighter – closer to each other – than in a horizon, the limit

$$\frac{m}{L} \leq \frac{c^2}{4G} \tag{20}$$

   arises for every physical system of size L. The limit has a value of $3.3666(1) \cdot 10^{26}\mathrm{kg/m}$ or about $1/6$ of a solar mass per km. Equality is predicted to hold *only for black holes.* The strand conjecture thus naturally implies that, for a given mass value, black holes are the densest objects in nature. Strands thus illustrate and imply both the *hoop conjecture* and the *Penrose conjecture:* for a given mass, because of the minimum size of crossings, a spherical horizon – a tight weave – has the smallest possible diameter. Other possible weave shapes have larger size. This agrees with expectations.

9. The strand conjecture illustrated in Figure 6 implies that black holes *evaporate*. Through fluctuations, single strands or tangles of strands can detach from the horizon weave. The strand conjecture allows deducing several predictions about evaporation.

   - First of all, the emission of particles will depend on the size of the black hole and on the tangling of the particle tangles, i.e., on particle mass values.
   - For large black holes, the evaporation is a low probability process, and the evaporation rate of such a black hole is small. This agrees with expectations.

     To calculate evaporation rates for different particles, probabilities for corresponding untangling processes must be calculated. At present, no mathematical tools to do this appear to exist. However, it is expected that particles made of one strand (photons) are emitted more frequently than particles made of two or three strands (gravitons and fermions). This agrees with all calculations [61].
   - The smaller the black hole, the higher the total luminosity, because strands detach with higher probability from a horizon with higher curvature.
   - For small black holes, the curvature of the black hole facilitates the emission of massive particle tangles. The relative probability for the emission of massive particles in black hole radiation is predicted to *increase* for smaller black holes, because for small black holes, the curvature helps emission of massive tangles.
   - Just before the completion of the evaporation process, black holes still radiate with a luminosity near but *below* the maximum possible value, the Planck power $c^5/4G$.

All these predictions agree with predictions made in the research literature decades ago [61].

10. Black holes evaporate until the horizon weave has completely *dissolved* into separate strands or tangles. Strands predict *the lack of black hole remnants* that differ from usual elementary particles.

11. Together with the strand description of black hole evaporation, strands predict and illustrate the lack of black holes with microscopic mass values. The corrected Planck limits for energy density, size, temperature and luminosity deduced above imply that all black holes obey

$$m > \sqrt{\frac{\hbar c}{4G}} \ , \tag{21}$$

thus have a mass that is *larger than the corrected Planck mass.* This agrees with observations and expectations.

12. The weave model of horizons also implies that elementary particles, which are tangles – not weaves – are *not* black holes. This agrees with expectations and with observations.

13. The strand conjecture automatically implies that the horizon area of a small black hole is *quantized* in multiples of the smallest area $4G\hbar/c^3$. This result was already deduced by Bekenstein [62]. However, strands also imply that area quantization of black holes is *not observable*, because in principle, no apparatus can have the sensitivity to detect this smallest area value. Such an apparatus would have to be able to count and thus to observe strands. This is impossible.

14. The strand conjecture for black holes of Figure 6 and the statistical properties of their fluctuations also imply that white holes do *not exist.* For reasons of probability, evaporation cannot take place backwards. This agrees with observations.

15. Because black hole horizons are weaves in the strand conjecture, black holes are predicted to have *no hair*, i.e., no nuclear charges, no baryon number, no lepton number or other quantum numbers. In previous papers [11, 15] it became clear that all these quantum numbers are topological properties of tangles. In the strand conjecture, these quantum numbers

are not defined for horizons. All quantum numbers except electric charge – which is defined with the help of crossing or tangle chirality and is explored below – do not make sense for weaves. The *no-hair theorem* is thus natural in the strand conjecture.

It is ironic that the strand conjecture can also be seen as a way to describe particles and horizons *only* with the help of "hair", if one uses "hair" as a synonym for "strand" or "tether". Using this terminology, one could say that the "hair conjecture" implies the no-hair theorem.

16. In the strand conjecture, horizons are tight, one-sided weaves. Any matter tangle that falls towards a horizon and is near it for a distant observer is flattened. As a result, at most one Planck mass can arrive at a horizon during a Planck time. Expressions (2) then yield the mass rate limit

$$\frac{\mathrm{d}m}{\mathrm{d}t} \leq \frac{c^3}{4G} \quad . \tag{22}$$

This limit – again valid for any point in space – is also valid in general relativity – and in nature in general. So far, the limit, given by $1.00928(3) \cdot 10^{35}\,\mathrm{kg/s}$ or $50\,756(12)$ solar masses per second, is not violated by any observation – including black hole mergers. Also, numerical simulations of general relativity did not exceed the limit [37].

17. In any physical system, strand crossings can be more or less tight, and switch more or less frequently. The limit case for a system of size R and energy E is the one with the *tightest* possible strands, as defined by the smallest length in expressions (2). This directly yields

$$\frac{2\pi}{\hbar c}\, ER \geq \frac{S}{k} \quad . \tag{23}$$

This is *Bekenstein's entropy bound.* The strand conjecture implies that equality is realized by horizons – and only by horizons – because horizons are the strand configurations that are as tight as possible and whose crossings switch as rapidly as possible. This agrees with expectations.

18. The strand conjecture *limits* energy density (and pressure) to the (corrected) Planck value:

$$\frac{E}{V} \leq \frac{c^7}{16\, G^2 \hbar} = 2.8958(1) \cdot 10^{112}\,\mathrm{J/m^3} \quad . \tag{24}$$

The energy density limit implies a lower limit for black hole size, for particle size and for the size of any localized system. Therefore, strands do *not allow mass singularities* in nature, neither dressed nor naked. Cosmic censorship is automatically realized in the strand conjecture. So far, both the density limit and the lack of mass singularities agree with observations.

19. As explained above, the strand conjecture for black holes of Figure 6 allows a further conclusion. For observers at rest outside the black hole, the weave model of horizons implies that *nothing* can be observed behind the horizon. In simple terms, nothing is 'inside' a black hole horizon. In particular, strands imply the *lack of a tightly concentrated mass* inside a black hole.

20. Strands imply that the mass of black holes is distributed over their horizon. Therefore, all black holes, including Schwarzschild black holes, have a finite *moment of inertia* I. Since strands reproduce general relativity, the moment of inertia of Schwarzschild black holes is given, as in general relativity, by the limit deduced for slowly rotating Kerr black holes, as determined by Raine [63] (page 68), by Ha [64, 65] and by Thorne et al. [66] (page 39):

$$I = MR^2 \ . \tag{25}$$

This result again disagrees with the idea that black hole mass is concentrated in a putative central singularity. Falsifying this value for the moment of inertia would falsify the strand conjecture.

The value of the moment of inertia is *larger* than that of a spherical mass shell, for which $I = 2MR^2/3$. The strand model visualizes the difference between a black hole and a mass shell in the following manner: Figure 8, showing the belt trick with a large number of belts, implies that every smallest surface on the horizon contributes the same number of crossing switches. Every smallest surface on the horizon thus contributes equally to the angular momentum, independently of its distance from the axis of rotation. Because mass is evenly distributed over the horizon, the total moment of inertia is $I = MR^2$.

21. In the strand conjecture, *electric charge* is a result of the *chiral* linking of strands [11, 15]. Because horizons are weaves of strands, the electric

charge Q of black hole horizons is *limited.* Strands visualize the limit directly.

A simple way to deduce the charge limit is to use the force limit $c^4/4G$. The electric force between two charged black holes must be smaller than the maximum force:

$$\frac{Q^2}{4\pi\epsilon_0\, R^2} \leq \frac{c^4}{4G} \; . \tag{26}$$

Using the black hole relation $M = Rc^2/2G$ of equation (3), this can be rewritten as

$$\frac{Q^2}{4\pi\epsilon_0} \leq GM^2 \; , \tag{27}$$

which is the established limit for a Reissner-Nordstrm black hole. Finding an exception to the charge limit would falsify the strand conjecture. However, such an exception would also falsify maximum force and general relativity. Unfortunately, no observations that allow testing the region near the limit are available so far. In fact, it is expected that virtual pair production prevents such observations.

22. Strands model black holes as weaves. Because strands model electric charge with crossing chirality, this implies that the electric charge of a black hole is distributed *over its surface.* The predicted charge distribution is consistent with the distribution of black hole *mass* mentioned above. Indeed, the strand conjecture implies that electric charge exists only for massive objects, and that charge and mass cannot be separated.

23. Being weaves, black holes can be either non-rotating or rotating. For rotating black holes, the strands in the weave provide a limit to the angular momentum of the black hole. Angular momentum, like spin, is a result of strand crossing switches [11, 15]. In a rotating black hole, the weave rotates. Because the equatorial speed is limited by c, a maximal rotation frequency ω arises, with the value $\omega \leq c/R$. Using the limit $J \leq E/\omega$, this yields

$$J \leq \frac{2G}{c} M^2 \; . \tag{28}$$

As expected, a rotating weave behaves like a Kerr black hole [67]. A higher angular momentum would contradict the fundamental principle, and in particular the minimum time for crossing switches. So far, the angular momentum limit for extremal black holes agrees with observations [68].

24. The *irreducible mass* of a rotating black hole is determined by the number of strands N_s that make it up. Strands thus predict that the *total* mass of a rotating black hole is a monotonous function of the irreducible mass and of its rotational energy, up to the angular momentum limit. This agrees with expectations and with observations.

25. The description of rotating black holes or masses with strands also suggests that moving tethers describe what is usually called *frame dragging*. In the strand conjecture and in general relativity, frame dragging occurs around all rotating masses, at all distances, and independently of whether the mass is a black hole or not. Like all other observable effects, also frame dragging results from crossing switches.

26. Strands also allow exploring black holes that are both charged and rotating – the Kerr-Newman case. In the strand conjecture, electric charge is due to the chirality of tangles [11, 15].

    Strands imply that when a charged *black* hole rotates, the tethers move as shown in Figure 8. This motion implies and predicts that the g-factor for such black holes is
    $$g = 2 \ . \qquad (29)$$
    Strands make this prediction (at tree level in the case of elementary particle) for all rotating systems for which the crossings that generate mass and those that generate charge rotate together. In all these cases, the g-factor is 2 because of the belt trick [11, 15]. The value 2 for the g-factor of black holes agrees with the usual predictions [69, 70, 71]. The question whether the g-factor is exactly 2 or whether it shows corrections that depend on the fine structure constant α – especially in the case of maximally charged black holes – is still open. So far, however, no way to test these predictions appears to be possible.

27. Strands allow expressing the results on rotating charged black holes with an additional limit. As deduced above, strands imply a charge limit for

any black hole given by equation (27). The definition of the g-factor

$$\mu = g\frac{Q}{2M}J \qquad (30)$$

implies, for $g = 2$, that

$$\frac{\mu}{J} = \frac{Q}{M}J \ . \qquad (31)$$

This means that

$$|\frac{\mu}{J}| \leq \sqrt{G}\sqrt{4\pi\epsilon_0} \ . \qquad (32)$$

Strands thus confirm the limit conjectured by Barrow and Gibbons [33]. So far, all observations and thought experiments agree with the limit.

In summary, the Barrow–Gibbons limit was derived from three strand properties: the horizon is a rotating weave; secondly, the electric charge, being due to chiral crossings, rotates with the mass; and thirdly, the crossings cannot rotate faster than the speed of light.

28. Strands predict the validity of the four laws of black hole mechanics [72]. The zeroth law, the constancy of surface gravity, is intrinsically valid in the weave model of horizons. The first law was deduced in Section 8.. The second law is realized by the detachment of strands during black hole evaporation. The third and last law is again automatic in the strand conjecture, as a result of the impossibility to reach the Planck limits. In fact, all four laws are built into the fundamental principle of the strand conjecture of Figure 3.

29. Strands predict that black holes have *vanishing* magnetic charge, because strands, due to their extension, do not allow magnetic charge to exist [11]. This agrees with expectations and observations so far.

30. Strands confirm that every horizon is a physical system that on the one hand can be seen as an extreme form of (curved) space, and on the other hand can be described as an extreme form of (falling) matter. Both points of view on horizons lead to tight, one-sided weaves as models for horizons. Horizons are thus systems at the border between space and matter. Alternatively, in the strand conjecture, *horizons are a mixture of space and matter.* This agrees with expectations.

31. The thermodynamic properties of strand fluctuations in black holes have implications for the shape oscillations of horizons. Shape oscillations of black hole horizons increase (and decrease) the local curvature. This increases (and decreases) the local evaporation through radiation, i.e., through strand detachment. As a result, horizon shape oscillations are *damped* and disappear over time. This agrees with theoretical expectations.

32. The strand conjecture for black holes illustrated in Figure 6 implies that for a distant observer at rest, horizons are not surfaces, but *thin cloudy volumes.* Strands thus imply that black hole horizons resemble stretched horizons. In contrast to an observer at rest, an observer *falling towards* and into the black hole experiences a three-dimensional strand network instead of an (almost) two-dimensional strand weave. The two descriptions can be transformed into each other with suitable deformations of the involved strands. The strand conjecture thus provides a model of a black hole that resembles a *'firewall'* [73] and a *'fuzzball'* [74, 75].

33. The strand conjecture implies that black holes (with all their quantum properties) are *impossible in higher dimensions*, because higher dimensions *do not allow* forming stable weaves. Strands thus imply that black holes can be imagined in higher dimensions only if quantum effects are (at least partially) neglected. However, this statement is hard or even impossible to verify.

34. The strand conjecture for black holes illustrated in Figure 6 suggests that black holes can *reflect* an incoming quantum particle, instead of swallowing it, but that the probability is *extremely* low: the incoming particle must have an energy so low that its wavelength is comparable to the size of the black hole. For such a low energy, the particle strands are similar in shape to vacuum strands, and the motion of the scattered particle around the black hole resembles the motion of vacuum strands around a traveling black hole. This low probability agrees with expectations [76].

11. Strand Predictions about Quantum Gravity and Gravitons

Strands predict the results of many quantum gravity experiments.

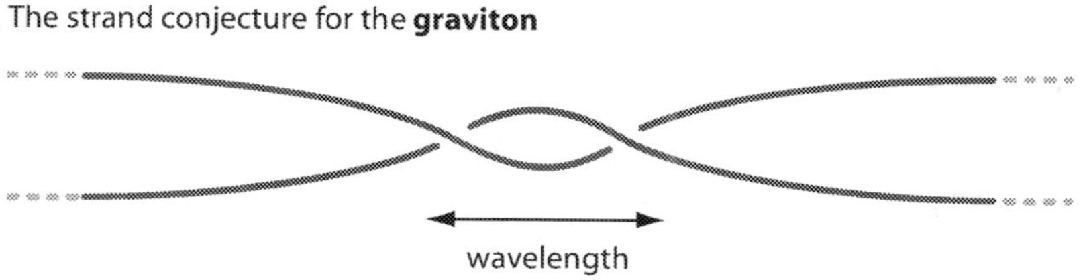

Figure 9. The strand conjecture for the graviton is a twisted pair of strands. The configuration has spin 2, boson behaviour, and zero mass. A macroscopic gravitational wave is an ensemble of a large number of gravitons. (The figure is modified from reference [12].)

35. Gravity is due to the exchange of virtual gravitons. The tangle model of the graviton, a twisted pair of strands that moves and rotates, is illustrated in Figure 9. Gravitons do surround masses: every Dirac trick generates twisted strand pairs. Strands also predict that gravitons have spin 2, because gravitons return to their original state after a rotation of the tangle core by π. Gravitons are predicted to be massless, because their core is not localized. Because cores can swap positions along the strands, gravitons are predicted to be bosons. As a result, coherent gravitons are predicted to yield gravitational waves with spin 2 and velocity c, as observed. Graviton exchange is also at the basis of universal $1/r^2$ gravity, as illustrated in Figure 7.

36. In the strand conjecture, *single gravitons cannot be detected*, for two reasons. First, strands imply the indistinguishability between graviton observation from any other quantum fluctuation of or at a detector. Equivalently, in the strand conjecture, the absorption of a single graviton does not lead to a detectable particle emission. Secondly, even if gravitons were detectable, in the strand conjecture, they have an extremely small cross section, of the order of the square of the Planck length. The low cross section is due to the graviton tangle shown in Figure 9. The tangle implies a low detection probability, as expected [77, 78].

37. Strands – and expressions (2) – imply that the gravitational constant G *does not run* with energy. In the language of perturbative quantum field theory, strands imply that G is *not renormalized*. This prediction agrees

with expectations [79, 80] and with observations, though the available data is sparse and many opposite views exist.

38. Strands imply that quantum superposition effects for gravitational systems are unobservable, because the exchange of graviton destroys entanglement. This is as expected [81].

    In particular, strands imply that in a *double-slit experiment* with quantum particles, the particles pass both slits at the same time; the particle core splits in two pieces during passage – though in different fractions at every passage. Therefore strands predict that the gravitational field of a quantum particle arises at both slits, at every passage, though each time in different fractions.

39. The impossibility to detect single gravitons implies the *lack* of unknown, observable *quantum corrections* to general relativity. Equivalently, strands predict the lack of observable quantum effects in *semiclassical gravity.*

So far, these predictions are not contradicted by any observation.

12. Strand Predictions about Elementary Particle Masses

The strand conjecture reproduces known physics. In addition, new results are possible [11, 15]. This section just mentions a few results about elementary particle masses.

Given that black holes are made of large numbers of woven strands, it is natural to assume that elementary particles are made of *a few* woven strands. Indeed, in the strand conjecture, all elementary particles are *rational tangles* – i.e., woven, unknotted tangles – made of one, two or three strands [11, 15]. Tangles made of four or more strands are composed, not elementary. An example of an elementary rational tangle is shown in Figure 10.

Among tangles made of a few strands, those made of one strand are bosons; more precisely, they are photons. Massive elementary particles tangles are made of two or three strands. Every fermion tangle, being a tethered structure that is tangled in the region of its tangle core, has non-vanishing mass. Every fermion tangle reproduces spin 1/2 behaviour under rotations – using Dirac's belt trick

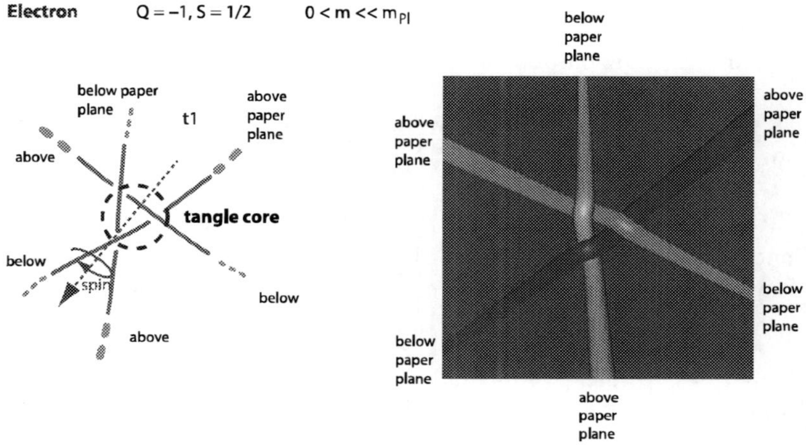

Figure 10. In the strand conjecture, elementary particles are modelled as rational tangles of strands. Tangles are called *rational* when they are formed by switching tethers. Tethers and strand segments are unobservable; only crossing switches are observable. Due to their fluctuations, tangles lead to the observation of particles that are localized in the region of the tangle core. Only rational tangles model the observed behaviour of elementary particles. Antiparticles are mirror tangles. The chirality determines the electric charge. Fermion tangles, such as the electron tangle shown in the figure, automatically have spin 1/2. Spin is core rotation. Wave functions are washed out tangle crossings; probability densities are washed out crossing switches. (The figure is modified from reference [12].)

– and fermion behaviour under the exchange of positions of tangle cores. All tangles reproduce the gauge groups U(1), SU(2) and SU(3) as the result of Reidemeister moves on their tangled cores.

Only *rational* tangles – i.e., tangles that arise through the motion or braiding of their tethers – allow reproducing the transformation of particles observed in experiments. And only rational tangles allow a classification into a finite number of families that correspond to the observed elementary particles. These arguments are worked out in detail in references [11] and [15].

Mass is the property of tangles that creates virtual gravitons around them. Mass is also given by action per time, divided by c^2. This implies:

▷ The *particle mass* (in corrected Planck units) is the probability of strand crossing switches occurring, per Planck time, in spontaneous belt tricks of the particle tangle.

Rational tangles directly allow deducing a number of predictions about mass values of elementary particles.

40. Strands promise, through the analogy between thermodynamic effects and gravitational attraction, to allow calculating the gravitational mass of quantum particles. The value of gravitational mass is predicted to depend on the geometric *tangle shape* of the particle – and on nothing else.

    Since particle mass is due to their (tight) tangle shape, the mass values of all elementary particles are predicted to be zero or positive, discrete but not quantized, equal to that of their antiparticles, fixed, unique, calculable and constant in time and space. This agrees with data. *If particle masses would be found to vary over space or time, the strand conjecture would be falsified.*

    In the strand conjecture, as shown in references [11, 15], only particles with positive mass can have electric and weak charge. In addition, it was shown that all mass values are due to Yukwawa coupling to the Higgs. Only those particles that couple to the Higgs are observed to be massive. All this agrees with observation.

41. The tangle model of elementary particles implies that both the gravitational and the inertial mass of elementary particles are due to tether fluctuations. *Gravitational* mass describes the virtual gravitons around a mass: they arise in the tethers due to the belt trick. *Inertial* mass describes how a rotating mass advances through the vacuum with the belt trick, as described by Battey-Pratt and Racey [16] and many others after them. In the strand conjecture, these two processes are exactly the same: both involve tether fluctuations around the core, and in particular, both involve the belt trick. Therefore, inertial and gravitational mass are equal – for infinite, flat space. Strands thus imply that the *equivalence principle* holds, in its weak and strong forms – at least for sub-galactic scales, when there is no effect of the cosmological horizon. This agrees with observations [82].

42. Strands imply that elementary particle mass values *run* with four-momentum. The reason is that the tangles completely reproduce quantum

field theory, and that elementary particles are surrounded by virtual particle pairs; thus their mass values run with four-momentum. This agrees with observations – e.g., [83] – and expectations.

43. Because spontaneous tangle fluctuations leading to the belt trick are *rare*, the gravitational mass m of elementary particles is to be much smaller than the corrected Planck mass:

$$0 < m \ll \sqrt{\hbar c/4G} \ . \tag{33}$$

This inequality agrees with observations and agrees with old arguments [84]. Strands thus provide a general answer to the *mass hierarchy* problem.

44. Strands imply that falling particles are fluctuating and diffusing tangles. This implies that *more complex* particle tangles have *higher* gravitational mass (for equal number of tethers). The same connection has already been deduced for inertial mass in a different way [11]. The connection yields the correct mass sequences for all hadrons and predicts normal mass ordering for neutrinos. *If neutrino masses would* not *obey normal ordering, the strand conjecture would be falsified.*

The tangle model also explains that neutrinos mix and that their mass values are stable under renormalization, as shown in references [11, 15]. Strands thus allow non-vanishing neutrino mass in the standard model of particle physics. More estimates about fermion masses using the tangle model are found in reference [12].

More precise strand estimates of particle masses require the development of better approximations and of suitable computer simulation programs. *The failure to reproduce the correct mass value of a single particle, at any single energy value, would falsify the strand conjecture.*

Conclusion

Thinking about nature as made of strands is unusual. It is also unusual to describe every physical process as a combination of fundamental events due to strand crossing switches. This unusual model for black hole horizons provides no new or unexpected results about black holes: strands predict all their classical and quantum properties. Due to this agreement, the strand model of black holes is interesting only for one reason.

Strands also allow deducing quantum field theory, the spectrum of forces, and the spectrum of elementary particles. In fact, strands predict the lack of any deviation from the standard model of particle physics, with massive neutrinos with PMNS mixing, as argued in detail elsewhere [11, 15]. As a new result,

45. Strands predict that the mass values of elementary particles, their coupling constants and their mixing angles, including their running with energy, can be calculated ab initio from tangle geometry.

Only a few predictions on particle masses have been given here; more predictions about mass and quantum gravity are found elsewhere [12]. As long as the predictions due to the strand conjecture are not falsified, the conjecture remains a candidate for a complete description of nature.

Acknowledgments and Declarations

The author thanks Jason Hise for his animations and Thomas Racey, Gary Gibbons, Arun Kenath, Chandra Sivaram, John Barrow, Eric Rawdon and Yuan Ha for discussions and suggestions.

References

[1] G. 't Hooft, *Int. J. Mod. Phys. A* **24**, 3243 (2009).

[2] L. Susskind, (2013), arXiv:1311.3335 [hep-th].

[3] C. Rovelli, **QGQGS2011**, 003 (2011), arXiv:1102.3660 [gr-qc].

[4] T. Padmanabhan, *AIP Conf. Proc.* **861**, 179 (2006), arXiv:astro-ph/0603114.

[5] T. Padmanabhan, *Rept. Prog. Phys.* **73**, 046901 (2010), arXiv:0911.5004 [gr-qc].

[6] T. Padmanabhan, *J. Phys. Conf. Ser.* **701**, 012018 (2016).

[7] H. Nicolai, *Fundam. Theor. Phys.* **177**, 369 (2014), arXiv:1301.5481 [gr-qc].

[8] G. Amelino-Camelia, *Living Rev. Rel.* **16**, 5 (2013), arXiv:0806.0339 [gr-qc].

[9] S. Carlip, *Class. Quant. Grav.* **34**, 193001 (2017), arXiv:1705.05417 [gr-qc].

[10] M. Botta Cantcheff, (2011), arXiv:1105.3658 [hep-th].

[11] C. Schiller, *Phys. Part. Nucl.* **50**, 259 (2019), arXiv:0905.3905 [physics.gen-ph].

[12] C. Schiller, *Indian Journal of Physics* 10.1007/s12648-021-02209-8 (2021a).

[13] N. Bohr, *Atomtheorie und Naturbeschreibung: Vier Aufsätze mit einer einleitenden Übersicht* (J. Springer, Berlin, Germany, 1931) p. 77.

[14] M. Gardner, *Riddles of the sphinx and other mathematical puzzle tales* (Mathematical Association of America, Washington, D.C, 1987).

[15] C. Schiller, *Eur. Phys. J. Plus* **136**, 79 (2021b).

[16] E. P. Battey-Pratt and T. J. Racey, *Int. J. Theor. Phys.* **19**, 437 (1980).

[17] E. A. Rauscher, *Lett. Nuovo Cim.* **7S2**, 361 (1973).

[18] H. J. Treder, *Found. Phys.* **15**, 161 (1985).

[19] R. J. Heaston, *Journal of the Washington Academy of Sciences* **80**, 25 (1990).

[20] V. de Sabbata and C. Sivaram, *Found. Phys. Lett.* **6**, 561 (1993).

[21] C. Massa, *Astrophys. Space Sci.* **232**, 143 (1995).

[22] L. Kostro and B. Lange, *Phys. Essays* **12**, 182 (1999).

[23] G. W. Gibbons, *Found. Phys.* **32**, 1891 (2002), arXiv:hep-th/0210109.

[24] C. Schiller, (2003), arXiv:0309118 [physics].

[25] C. Schiller, *Int. J. Theor. Phys.* **44**, 1629 (2005), arXiv:physics/0607090.

[26] C. Schiller, *Int. J. Theor. Phys.* **45**, 221 (2006a).

[27] J. Barrow and G. Gibbons, *Mon. Not. Roy. Astron. Soc.* **446**, 3874 (2015).

[28] M. P. Dabrowski and H. Gohar, *Phys. Lett. B* **748**, 428 (2015), arXiv:1504.01547 [gr-qc].

[29] M. R. R. Good and Y. C. Ong, *Phys. Rev. D* **91**, 044031 (2015), arXiv:1412.5432 [gr-qc].

[30] Y. L. Bolotin, V. A. Cherkaskiy, A. V. Tur, and V. V. Yanovsky, (2016), arXiv:1604.01945 [gr-qc] .

[31] V. Cardoso, T. Ikeda, C. J. Moore, and C.-M. Yoo, *Phys. Rev. D* **97**, 084013 (2018), arXiv:1803.03271 [gr-qc].

[32] Y. C. Ong, *Phys. Lett. B* **785**, 217 (2018), arXiv:1809.00442 [gr-qc].

[33] J. D. Barrow, *Class. Quant. Grav.* **37**, 125007 (2020a), arXiv:2002.10155 [gr-qc].

[34] J. D. Barrow, *Int. J. Mod. Phys. D* **29**, 2043008 (2020b), arXiv:2005.06809 [gr-qc].

[35] J. D. Barrow and N. Dadhich, *Phys. Rev. D* **102**, 064018 (2020), arXiv:2006.07338 [gr-qc].

[36] K. Atazadeh, *Phys. Lett. B* **820**, 136590 (2021).

[37] L.-M. Cao, L.-Y. Li, and L.-B. Wu, *Phys. Rev. D* **104**, 124017 (2021), arXiv:2109.05973 [gr-qc].

[38] A. Jowsey and M. Visser, Universe **7** (2021), arXiv:2102.01831 [gr-qc].

[39] V. Faraoni, *Phys. Rev. D* **103**, 124010 (2021a), arXiv:2105.07929 [gr-qc].

[40] C. Schiller, *Phys. Rev. D* **104**, 068501 (2021c), arXiv:2109.07700 [gr-qc].

[41] V. Faraoni, *Phys. Rev. D* **104**, 068502 (2021b).

[42] C. Schiller, *Phys. Rev. D* **104**, 124079 (2021d), arXiv:2112.15418 [gr-qc].

[43] C. Sivaram, A. Kenath, and C. Schiller, preprint (2021).

[44] J. D. Bekenstein, *Phys. Rev. D* **7**, 2333 (1973).

[45] W. H. Zurek and K. S. Thorne, *Phys. Rev. Lett.* **54**, 2171 (1985).

[46] S. Carlip, *Class. Quant. Grav.* **17**, 4175 (2000), arXiv:gr-qc/0005017.

[47] L. Smarr, *Phys. Rev. Lett.* **30**, 71 (1973), [Erratum: Phys. Rev. Lett. 30, 521–521 (1973)].

[48] T. Jacobson, *Phys. Rev. Lett.* **75**, 1260 (1995), arXiv:gr-qc/9504004. NoStop

[49] A. Raychaudhuri, *Phys. Rev.* **98**, 1123 (1955).

[50] D. Christodoulou and S. Klainerman, *The global nonlinear stability of the Minkowski space* (1993).

[51] J. W. Richardson, O. Kwon, H. R. Gustafson, C. Hogan, B. L. Kamai, L. P. McCuller, S. S. Meyer, C. Stoughton, R. E. Tomlin, and R. Weiss, *Phys. Rev. Lett.* **126**, 241301 (2021), arXiv:2012.06939 [gr-qc].

[52] R. Cooke, L. Welsh, M. Fumagalli, and M. Pettini, *Mon. Not. Roy. Astron. Soc.* **494**, 4884 (2020), arXiv:2001.06016 [astro-ph.CO].

[53] P. Laurent, D. Gotz, P. Binetruy, S. Covino, and A. Fernandez-Soto, *Phys. Rev. D* **83**, 121301 (2011), arXiv:1106.1068 [astro-ph.HE].

[54] A. H. Chamseddine (2018) arXiv:1805.08582 [hep-th].

[55] A. H. Chamseddine, A. Connes, and M. Marcolli, *Adv. Theor. Math. Phys.* **11**, 991 (2007), arXiv:hep-th/0610241.

[56] T. P. Singh, *Zeitschrift für Naturforschung A* **75**, 833 (2020).

[57] R. Penrose, *Reports on Mathematical Physics* **12**, 65 (1977).

[58] C. Schiller, *International Journal of Theoretical Physics* **45**, 213 (2006b).

[59] B. P. Abbott et al. (LIGO Scientific, Virgo), *Phys. Rev. X* **9**, 031040 (2019), arXiv:1811.12907 [astro-ph.HE].

[60] R. Abbott et al. (LIGO Scientific, Virgo), *Astrophys. J. Lett.* **900**, L13 (2020), arXiv:2009.01190 [astro-ph.HE].

[61] V. Frolov and I. Novikov, *Black hole physics: Basic concepts and new developments*, Vol. 96 (Springer Science & Business Media, 2012).

[62] J. D. Bekenstein, in *9th Brazilian School of Cosmology and Gravitation* (1998) arXiv:gr-qc/9808028.

[63] D. Raine and E. Thomas, *Black Holes: An Introduction*, Black Holes: An Introduction (Imperial College Press, 2010).

[64] Y. K. Ha, *Int. J. Mod. Phys. D* **14**, 2219 (2005), arXiv:gr-qc/0509063 .

[65] Y. K. Ha, *Int. J. Mod. Phys.*A **33**, 1844025 (2018), arXiv:1811.02890 [physics.gen-ph] .

[66] K. S. Thorne, R. H. Price, and D. A. Macdonald, eds., *Black Holes: The Membrane Paradigm* (1986).

[67] S. Dain and M. E. Gabach-Clement, Living Rev. Rel. **21**, 5 (2018), arXiv:1710.04457 [gr-qc].

[68] C. S. Reynolds, *Nature Astron.* **3**, 41 (2019), arXiv:1903.11704 [astro-ph.HE].

[69] R. M. Wald, Phys. Rev. D **10**, 1680 (1974).

[70] H. Pfister and M. King, *Classical and Quantum Gravity* **20**, 205 (2002).

[71] B. R. Holstein, *Am. J. Phys.* **74**, 1104 (2006), arXiv:hep-ph/0607187.

[72] J. M. Bardeen, B. Carter, and S. W. Hawking, *Commun. Math. Phys.* **31**, 161 (1973).

[73] A. Almheiri, D. Marolf, J. Polchinski, and J. Sully, *Journal of High Energy Physics* **2013**, 62 (2013).

[74] S. D. Mathur, (2015), arXiv:1506.04342 [hep-th].

[75] S. D. Mathur, (2014), arXiv:1401.4097 [hep-th].

[76] M. Y. Kuchiev, *Phys. Rev. D* **69**, 124031 (2004), arXiv:gr-qc/0310051.

[77] T. Rothman and S. Boughn, *Found. Phys.* **36**, 1801 (2006), arXiv:gr-qc/0601043.

[78] F. Dyson, *Int. J. Mod. Phys. A* **28**, 1330041 (2013).

[79] M. M. Anber and J. F. Donoghue, *Phys. Rev. D* **85**, 104016 (2012), arXiv:1111.2875 [hep-th].

[80] J. F. Donoghue, *Front. in Phys.* **8**, 56 (2020), arXiv:1911.02967 [hep-th].

[81] S. Rijavec, M. Carlesso, A. Bassi, V. Vedral, and C. Marletto, *New J. Phys.* **23**, 043040 (2021), arXiv:2012.06230 [quant-ph].

[82] C. M. Will, *Living Rev. Rel.* **17**, 4 (2014), arXiv:1403.7377 [gr-qc] .

[83] A. M. Sirunyan et al. (CMS), *Phys. Lett. B* **803**, 135263 (2020), arXiv:1909.09193 [hep-ex].

[84] V. A. Berezin, *Phys. Part. Nucl.* **29**, 274 (1998).

Chapter 6

From Bohr to Schrodinger: Two-Particle System Approach to Black Hole Quantum Gravity

Christian Corda*
International Institute for Applicable Mathematics and Information Sciences,
B. M. Birla Science Centre, Adarshnagar, Hyderabad, India
and Istituto Livi, Prato, Italy

Abstract

A novel approach to black hole (BH) quantum gravity which has been recently developed by the Author and Collaborators in a series of research papers is carefully reviewed. In such an approach the Schwarzschild BH (SBH) emerges as a quantum non-singular two-particle system where the two components, the "nucleus" and the "electron", strongly interact with each other through a quantum gravitational interaction. After summarizing the general quantum gravity problem, the analysis starts from the semi-classical Bohr-like approach to BH quantum physics which was early developed by the Author and Collaborators. After that, by applying both Rosen's quantization approach and Feynman's path integrals framework to the historical Oppenheimer and Snyder gravitational collapse and by setting the constraints for the formation of the SBH, the gravitational potential, the Schrodinger equation and the solution for the energy levels will be derived. Remarkably, such results are consistent with previous

*Corresponding Author's Email: cordac.galilei@gmail.com.

In: A Guide to Black Holes
Editor: Kenath Arun
ISBN: 979-8-88697-163-7

ones in the literature. Hence, the obtaied results show that the SBH is the gravitational analogous of the hydrogen atom and this could, in principle, drive to a space-time quantization based on a quantum mechanical particle approach. Following this analogy with the hydrogen atom, one can consider the de Broglie hypothesis and the wave nature of the BH "electron". This means that such a particle does not orbit the nucleus in the same way as a planet orbits the Sun, but instead exists as a standing wave. The correct analogy is that of a large and often oddly shaped "atmosphere" (the BH "electron"), distributed around a relatively tiny planet (the BH "nucleus"). The correct physical interpretation of such an "atmosphere" is nothing else than the BH horizon modes. By following with caution the analogy between the SBH Schrodinger equation and the traditional Schrodinger equation of the s states ($l = 0$) of the hydrogen atom, the SBH Schrodinger equation is solved and discussed in detail. The approach also permits us to find the quantum gravitational quantities which are the gravitational analogous of the fine structure constant and of the Rydberg constant. Remarkably, such quantities are not constants. Instead, they are dynamical quantities having well defined discrete spectra. In particular, the spectrum of the "gravitational fine structure constant" is exactly the set of non-zero natural numbers $\mathbb{N} - \{0\}$. Therefore, one argues the interesting consequence that the SBH results in a well defined quantum gravitational system, which obeys Schrodinger's theory: the "gravitational hydrogen atom". The area quantum and the quantum representation of the SBH ground state at the Planck scale will be also derived. Finally, it is shown that, remarkably, the time evolution of the "gravitational hydrogen atom", which is governed by a time-dependent Schrodinger equation, solves the BH information paradox.

1. Introduction: The Quantum Gravity Problem and the Importance of the Quantum Black Hole

It is often emphasized by theoretical physicists that the two main pillars of modern physics are Einstein's general relativity and quantum theory. On the one hand general relativity well-describes gravitational phenomena at large scales, starting from observations from cosmological distances to millimeter scales. On the other hand, quantum mechanics and quantum field theory well-describe phenomena at small scales from a fraction of a millimeter down to 10^{-19} meters. Such scales are dominated by strong and electroweak interactions. Although general relativity [1] achieved great success (see for example the opinion of Landau who says that general relativity is, together with quantum field theory, the best scientific theory of all [2]) and withstood many experimental tests, it also displayed many shortcomings and flaws which today make theoreticians question whether it is the definitive theory of gravity [3]. As distinct from other field theories, like the electromagnetic theory, general relativity is very difficult to quantize. This fact rules out the possibility of treating gravitation like other quantum theories and precludes the unification of gravity with other interactions. At the present time, it is not possible to realize a consistent *quantum gravity theory* which leads to the unification of gravitation with the other forces. From a historical point of view, Einstein believed that, in the path to unification of theories, quantum mechanics had to be subjected to a more general deterministic theory, which he called *generalized theory of gravitation*, but he did not obtain the final equations of such a theory (see for example the biography of Einstein which has been written by Pais [4]). At present, this point of view is partially retrieved by some theorists [5]. During the last 50 years, a strong, critical discussion about both quantum mechanics [6, 8] and general relativity [7] has been undertaken by theoreticians in the Scientific Community. Details on the general discussion on the quantum gravity problem can be found in [5]. Here one limits himself to discuss which is the BHs' role in the framework of quantum gravity. It is a general conviction, which arises from the famous, pioneering works of Bekenstein [9] and Hawking [10], that the role and the importance of BHs are fundamental. BHs are indeed considered as being theoretical laboratories for testing different models of quantum gravity. Bekenstein was the first physicist who observed that, in some respects, BHs play the same role in gravitation that atoms played in the nascent quantum mechanics [11]. This analogy implies that BH energy could have a discrete spectrum [11]. Therefore, BHs combine in some sense both the "hydrogen atom" and the "quasi-thermal" emission in quantum gravity [12]. As a consequence, BH quantization could be the key to a quantum theory of gravity and, for that reason the study of quantum BHs became, and currently remains, one of the most important research fields in theoretical physics of the last 50 years. Various Authors proposed and still propose various different approaches. Hence, the current literature is very rich, see for example [13–26] and references within. In this framework, the Author developed a semi-classical approach to BH quantization [27–36], also with some international collaborations [37–42], which is somewhat similar to the historical semi-classical approach to the structure of a hydrogen atom introduced by Bohr in 1913 [43, 44]. Recently, the Author and Collaborators [45–48] improved the analysis by adapting an approach to quantization of the famous collaborator of

Einstein, N. Rosen [49] to the historical Oppenheimer and Snyder gravitational collapse [50]. The approach in [45–48] has shown that the traditional classical singularity in the core of the SBH is replaced by a nonsingular two-particle system where the two components, the "nucleus" and the "electron", strongly interact with each other through a quantum gravitational interaction. In other words, the SBH is the gravitational analogous of the hydrogen atom, and this could, in principle, drive to a space-time quantization based on a quantum mechanical two-particle system approach. One asks, what is the physical meaning of the SBH "electron"? Following the analogy with the hydrogen atom, one can consider the de Broglie hypothesis [51] and the wave nature of the BH "electron". This means that such a particle does not orbit the nucleus in the same way as a planet orbits the Sun, but instead exists as a standing wave. The correct analogy is that of a large and often oddly shaped "atmosphere" (the BH "electron"), distributed around a relatively tiny planet (the BH "nucleus"). The correct physical interpretation of such an "atmosphere" is nothing else than the BH horizon modes. In fact, the idea that the radius of the event horizon undergoes quantum oscillations has a longstanding history. Such horizon modes were introduced in a semi-classical framework about 50 years ago [52] in terms of BH quasi-normal modes (QNMs) which represent the BH back reaction to perturbations. Both of the absorptions of external particles and the emissions of Hawking quanta are BH perturbations and this allowed the Author and Collaborators to develop the Bohr-like approach to BH quantum physics [27–42] starting from a very interesting paper written by Hod [12] and then improved by Maggiore [53]. On one hand, the QNMs approach is a semi-classical approach similar to the approach that Bohr developed in 1913 [43, 44] concerning the structure of the hydrogen atom. On the other hand, the importance of horizon modes in a quantum gravity framework has been recently emphasized in [54], by considering them as being described by the periodic motion of their particle-like analogue, in full accordance with the de Broglie hypothesis. The Authors of [54] found an energy spectrum which scales as $\sim \sqrt{n}$, consistent with the results obtained in [27–42]. The key point is that during both the processes of absorptions of external particles, including the original BH formation, and of emissions of Hawking quanta, the BH horizon is not fixed at a constant distance from the BH "nucleus" [32]. In fact, because of energy conservation, the BH contracts during the emission of a particle and expands during an absorption [32]. Such quantum contractions/expansions are not "one shot processes" [32]. They generate oscillations of the horizon instead [32].

The Book Chapter is organized as it follows. In Section 2, the Bohr-like approach to BH quantum physics developed by the Author and collaborators in [27–42] will be carefully reanalysed and refined. In Section 3 the Schrodinger theory of the SBH will be developed. In Subsection 3.1 the historical Oppenheimer and Snyder gravitational collapse [50] will be quantized via Feynman's path integrals approach. This will permit quantization of the SBH in Subsection 3.2 by applying the constraints for the BH formation. It will be shown that it is exactly the quantization procedure which splits the classical SBH in a two-particle system which consists in a "gravitational hydrogen atom". It will be also shown that the semi-classical Bohr-like approach to BH quantum physics, developed in [27–42] and reviewed and refined in Section 2, is consistent with the obtained results for large values of the BH principal quantum number. In Subsection 3.3 it will be shown that, remarkably, the time evolution of the "gravitational hydrogen atom" is governed by a time-dependent Schrodinger equation generating an unitary evolution which solves the BH information paradox.

Finally, Section 4 is dedicated to the conclusion remarks.

2. Bohr-Like Approach to Black Hole Quantum Physics

In this Section the Bohr-like approach to BH quantum physics which has been developed by the Author and Collaborators in [27–42] will be reviewed and refined. Let us start by recalling that, analyzing Hawking Radiation [10] as tunneling, Parikh and Wilczek showed that the radiation spectrum cannot be strictly thermal [55, 56]. An intriguing physical interpretation of this important issue has been proposed by Parikh [55], by discussing the existence of a secret tunnel through the BH horizon. The energy conservation implies that the BH contracts during the process of radiation [55, 56]. Thus, the horizon recedes from its original radius to a new, smaller radius [55, 56]. The consequence is that BHs cannot strictly emit thermally [55, 56]. This is consistent with the unitarity of BH evaporation and has important implications for the BH information paradox because arguments that information is lost during BH evaporation rely in part on the assumption of strict thermal behavior of the spectrum [57].

Working with $G = c = k_B = \hbar = \frac{1}{4\pi\epsilon_0} = 1$ (Planck units), the probability of emission is [10]

$$\Gamma \sim \exp(-\frac{\omega}{T_H}), \tag{1}$$

where $T_H \equiv \frac{1}{8\pi M}$ is the Hawking temperature and ω the energy-frequency of the emitted radiation. Parikh and Wilczek released a remarkable correction, due to an exact calculation of the action for a tunneling spherically symmetric particle, which yields [55, 56]

$$\Gamma \sim \exp[-\frac{\omega}{T_H}(1-\frac{\omega}{2M})]. \tag{2}$$

This result is in contrast with the one in [10] because it releases the additional term $\frac{\omega}{2M}$ as a deviation from the strict thermality [55, 56]. The key point is that in [10] the dynamical BH geometry due to the energy conservation has not been taken into due account by Hawking like by Parikh and Wilczek in [55, 56]. The energy conservation forces indeed the BH to contract during the emission of the particle [55, 56]. In this way, the horizon recedes from its original radius and becomes smaller at the end of the emission process [55, 56]. Thus, BHs do not exactly emit like perfect black bodies [55, 56]. In fact, the tunnelling is a *discrete* instead of *continuous* process [32, 33] because two different *countable* BH physical states have to be considered, the first before the emission of the particle and the latter after the emission of the particle [32, 33]. Then, the emission of the particle is interpreted like a *quantum transition* of frequency ω between the two different discrete states [32, 33]. The tunnelling mechanism works indeed considering a trajectory in imaginary or complex time joining two separated classical turning points [55, 56]. The important consequence is that the radiation spectrum is also discrete [32, 33]. Based on its importance, this issue has to be clarified in a better way. If one fixes the Hawking temperature, the statistical probability distribution (2) is a continuous function. But the Hawking temperature in (2) varies in time with a behavior which is *discrete.* The reason for this discrete character is that the forbidden region traversed by the emitting particle has a *finite* size [32, 33, 55, 56]. Considering a strictly thermal approximation, the turning points have zero separation. Thus, it is not clear what joining trajectory one has to consider because there is no barrier [55]. The problem is solved if one argues that it is the forbidden finite region from $r_{initial} = 2M$ to $r_{final} = 2(M-\omega)$ that the tunnelling particle traverses which acts like barrier [55]. Then, the intriguing explanation is that it is the particle itself which generates a tunnel through the BH horizon [55]. If one wants to take into due account the dynamical geometry of the BH during the emission of the particle a BH *effective state* can be introduced [27–42]. In fact, introducing the *effective temperature* [27–42]

$$T_E(\omega) \equiv \frac{2M}{2M-\omega}T_H = \frac{1}{4\pi(2M-\omega)}, \tag{3}$$

Eq. (2) can be rewritten in a Hawking-Boltzmann-like form similar to the original probability found by Hawking

$$\Gamma \sim \exp[-\beta_E(\omega)\omega] = \exp(-\frac{\omega}{T_E(\omega)}), \tag{4}$$

being $\exp[-\beta_E(\omega)\omega]$ the *effective Boltzmann factor* with [31, 32, 33]

$$\beta_E(\omega) \equiv \frac{1}{T_E(\omega)}. \tag{5}$$

Therefore, the effective temperature replaces the Hawking temperature in the equation of the probability of emission [27–42]. Let us discuss the physical interpretation of $T_E(\omega)$ following [31, 32, 33]. The probability of emission of Hawking quanta found in [55, 56], i.e. Eq. (2), shows that the BH does NOT emit like a perfect black body, having not a strictly thermal behavior. On the other hand, the temperature in Bose-Einstein and Fermi-Dirac distributions is a perfect black body temperature. Thus, when one has deviations from the strictly thermal behavior, i.e., from the perfect black body, one expects also deviations from Bose-Einstein and Fermi-Dirac distributions [58]. How can one attack this problem? By analogy with other various fields of Science, also beyond BHs, for example the case of planets and stars [31, 33, 59]. One defines the effective temperature of a body such as a star or planet as the temperature of a black body that would emit the same total amount of electromagnetic radiation [31, 33, 59]. The importance of the effective temperature in a star is stressed by the issue that the effective temperature and the bolometric luminosity are the two fundamental physical parameters needed to place a star on the Hertzsprung–Russell diagram [31, 33, 59]. Both effective temperature and bolometric luminosity actually depend on the chemical composition of a star [31, 33, 59]. The concept of effective temperature has been introduced by the author in BH physics in [27, 28] for the Schwarzschild BH and generalized in [37] to the Kerr BH and in [38] to the non-extremal Reissner-Nordström BH by the author and collaborators. $T_E(\omega)$ depends on the energy-frequency of the emitted radiation while the ratio $\frac{T_E(\omega)}{T_H} = \frac{2M}{2M-\omega}$ represents the deviation of the BH radiation spectrum from the strictly thermal behavior. After the introduction of $T_E(\omega)$ one can introduce other *effective quantities*. Considering the initial BH mass *before* the emission, M, and the final BH mass *after* the emission, $M - \omega$ respectively [31, 32, 33], the *effective BH mass* and the *effective BH horizon during* its contraction, i.e. *during* the emission of the particle are defined as [31, 32, 33]

$$M_E \equiv M - \frac{\omega}{2}, \; r_E \equiv 2M_E. \tag{6}$$

These effective quantities are average quantities [31, 32, 33]. The effective horizon r_E is the average of the initial and final horizons and the effective mass M_E is the average of the initial and final masses [31, 32, 33]. *Before* the emission the Hawking temperature is $T_{H \text{ initial}} = \frac{1}{8\pi M}$; *after* the emission one gets $T_{H \text{ final}} = \frac{1}{8\pi(M-\omega)}$. Then, T_E results to be the inverse of the average value of the inverses of the initial and final Hawking temperatures [31, 32, 33]. This implies that the Hawking temperature *has a discrete character in time* [31, 32, 33]. One notes that the introduction of the effective temperature does not degrade the importance of the Hawking temperature [31, 32, 33]. In fact, as T_H changes with a discrete behavior in time, let us ask: which value of the Hawking temperature has to be associated with the emission of the particle? Has one to consider the value of the Hawking temperature *before* the *emission* or the value of the Hawking temperature after the emission? The answer is that one has to consider an *intermediate* value, the effective temperature, which is the inverse of the average value of the inverses of the initial and final Hawking temperatures [31, 32, 33]. In some way, $T_E(\omega)$ represents the value of the Hawking temperature *during* the emission of the quantum [31, 32, 33]. Then, the effective temperature takes into account the non-strictly thermal character of the radiation spectrum and the non-strictly continuous character of subsequent emissions of Hawking quanta.

Now it will be shown that the correction to the thermal spectrum in [55, 56] is also very important for the physical interpretation of the quantum oscillations of the BH horizon, which in a semi-classical approximation are the QNMs, "triggered" by the emissions of Hawking quanta and by the absorptions of external particles [31, 32, 33]. The intriguing idea that BH QNMs carry important information about BH quantization is due to Hod [12]. Hod's original proposal was refined by Maggiore [53], who solved some problems. The Author and Collaborators further improved Maggiore analysis by realizing the Bohr-like approach to BH quantum physics [27–42] where the QNMs "triggered" by the emissions of Hawking quanta and by the absorptions of external particles are naturally interpreted in terms of quantum levels, the "electron states" of the SBH [31, 32, 33].

In general, BH QNMs are modes of radial perturbations obeying a time independent Schröedinger-like equation, see [27, 28, 29, 33, 62] for details. The intriguing idea to model the quantum BH in terms of QNMs was pioneered by York in [63]. By using the statistical mechanics of the QNMs, York obtained the value of $((0.27654))\left(16\pi M^2\right)$ for the BH entropy, which is near the value usually assumed of the standard Bekenstein-Hawking entropy $4\pi M^2$. Considering *Bohr's Correspondence Principle* [64], which states that "*transition frequencies at large quantum numbers should equal classical oscillation frequencies*", Hod found the important result that QNMs release information about the area quantization [12]. In his approach Hod considered the QNMs as associated with absorption of particles [12]. Some important problems arising from this approach were solved by Maggiore [53]. A fundamental issue is that QNMs are *countable* frequencies. This was in contrast with ideas on the *strictly thermal (non-countable)* character of Hawking radiation, preventing attempts to interpret QNMs in terms of emitted quanta, and, in turn, to associate QNMs to Hawking radiation [62]. On the other hand, the authors of [65–68] and ourselves and Collaborators [27–42] observed that the non-thermal spectrum in [55, 56] also implies the countable character of subsequent emissions of Hawking quanta. This key point enables a natural correspondence between Hawking radiation and BH QNMs [27–42]. Consequently, QNMs can be interpreted also in terms of emitted energies [27–42]. In fact, BH QNMs represent the BH reaction to small, discrete perturbations in terms of damped oscillations [27–42]. If the capture of a particle causing an increase in the horizon area is a type of discrete perturbation [12, 53, 62], it is obvious and natural to consider the emission of a particle causing a decrease in the horizon area also as a perturbation which generates a reaction in terms of countable QNMs, being that process countable instead of non-countable [27–42]. This can be immediately understood if one considers that it is the particle having negative energy and tunnelling inwards which is captured by the hole in Hawking's mechanism of pair creation. Thus, the absorbed Hawking quanta having negative energies generate subsequent perturbations "triggering" the QNMs. Concerning this key point, it is important to emphasize that the correspondence existing between emitted radiation and proper oscillation of the emitting body is well known as a fundamental behavior of every radiation process in Science. This natural correspondence between Hawking radiation and BH QNMs, permits immediately and naturally to interpret BH QNMs in terms of quantum levels also for emitted energies [27–42]. This also agrees with the below stressed general idea that BHs can be considered in terms of highly excited states representing both the "hydrogen atom" and the "quasi-thermal emission" in an underlying theory of quantum gravity.

Considering a strictly thermal approximation, BH QNMs are usually labelled as ω_{nl}, being l the angular momentum quantum number [12, 53, 62]. For each l, one finds a countable sequence of BH QNMs, labelled by the principal quantum number n ($n = 1, 2, ...$) [12, 53, 62]. For large n the Schwarzschild BH QNMs result independent of l. They have the structure [12, 53, 62, 69, 70, 71]

$$\omega_n \simeq \frac{2\pi i n}{8\pi M} \tag{7}$$

In fact, the leading term in the imaginary part of the complex frequencies becomes dominant [12, 33, 53, 62], and this is consistent with Bohr correspondence principle [64]. The Author of [62] has shown that Eq. (7) holds also for vector and half integer spin perturbations, again in agreement with Bohr correspondence principle. A key point is that Eq. (7) works only in strictly thermal approximation [27–42]. If one wants to take into due account the deviation from the perfect black body spectrum the Hawking temperature T_H must be replaced by the effective temperature T_E in Eq. (7) obtaining [27–42]

$$\omega_n \simeq \frac{2\pi i n}{4\pi\left[2M - |\omega_n|\right]}. \tag{8}$$

The Author released a first, intuitive derivation of Eq. (8) in [27, 28] and a more rigorous one in the Appendix of [29]. Further details on that derivation can be analysed following [33]. The BH *dynamical* geometry during the emission of the particle is taken into account by replacing T_H with T_E in Eq. (7). The passage from Eq. (7) to Eq. (8) has been rigorously justified in [29, 33] via Hawking's periodicy argument [60]. Eq. (8) has the following elegant interpretation [27, 28, 33]. QNMs determine the position of poles of a Green's function on the given background and the Euclidean BH solution converges to a *non-strictly* thermal circle at infinity with the inverse temperature $\beta_E(\omega_n) = \frac{1}{T_E(|\omega_n|)}$ [27, 28, 33]. Then, the spacing of the poles in Eq. (8) coincides with the spacing $2\pi i T_E(|\omega_n|) = 2\pi i T_H(\frac{2M}{2M-|\omega_n|})$, expected for a *non-strictly* thermal Green's function [27, 28, 33]. At order of leading asymptotic behavior the solution of Eq. (8) is [27, 28, 33]

$$|\omega_n| = M - \sqrt{M^2 - \frac{n}{2}}. \tag{9}$$

In [27, 28, 33] $|\omega_n|$ has been interpreted like the total energy emitted for a BH excited at a level n. Now, if one considers an emission from a BH at rest to a state with large $n = n_1$ and using Eq. (9), one sees that the BH mass changes from M to [27, 28, 33]

$$M_{n_1} \equiv M - |\omega_{n_1}| = \sqrt{M^2 - \frac{n_1}{2}}. \tag{10}$$

One considers a subsequent emission from the quantum level $n = n_1$ to the quantum level $n = n_2$ with $n_2 > n_1$. Then, the BH mass changes again from M_{n_1} to [27, 28, 33]

$$\begin{aligned} M_{n_2} \equiv M_{n_1} - \Delta E_{n_1 \to n_2} = M - |\omega_{n_2}| \\ = \sqrt{M^2 - \frac{n_2}{2}}, \end{aligned} \tag{11}$$

with

$$\Delta E_{n_1 \to n_2} \equiv |\omega_{n_2}| - |\omega_{n_1}| = M_{n_1} - M_{n_2} = \sqrt{M^2 - \frac{n_1}{2}} - \sqrt{M^2 - \frac{n_2}{2}}, \tag{12}$$

is the quantum jump between the two BH energy levels due to the emission of a particle having frequency $\Delta E_{n_1 \to n_2}$. Hence, the quantum jump implies the emission of a discrete amount of energy. As one considers large values of n, the analysis is independent of the other quantum numbers. Thus, one finds again a strong consistency with Bohr's Correspondence Principle [64]. In the analogy with Bohr hydrogen atom [43, 44], QNMs are considered the "electron" which jumps from a level to another one and the absolute values of the QNMs frequencies are considered the energy "shells" [27, 28, 33]. In the Bohr hydrogen atom [43, 44] the electron only gains and loses energy when it jumps from one allowed energy shell to another. It can absorb or emit radiation having an energy difference of the levels governed by the Planck relation (in standard units) $E = hf$, being h the Planck constant and f the transition frequency. In the Bohr-like approach to BH physics, QNMs can only gain and lose energy by jumping from one allowed energy shell to another [27, 28, 33]. Thus, they absorb or emit radiation (emitted radiation is given by Hawking quanta) with an energy difference of the levels given by to Eq. (12). Remarkably, one interprets Eq. (9) in terms of a particle, the "electron", quantized on a circle of length [27, 28, 33]

$$L = \frac{1}{T_E(|\omega_n|)} = 4\pi \left(M + \sqrt{M^2 - \frac{n}{2}} \right). \tag{13}$$

This is analogous to the electron travelling in circular orbits around the hydrogen nucleus, similar in structure to the solar system, of the Bohr approach [43, 44]. Thus, the cited analogy is complete.

Now, following [48] the semi-classical Bohr-like approach to BH physics will be refined through some simple, but important physical observations, and this will permit to find a perfect consistence with Feynman's full quantization approaches which will be developed in next Section. One starts with the observation that the Bohr-like approach has been developed starting from the assumption of a SBH at rest which starts to emit Hawking radiation. But the key point is that a BH can NEVER be at rest. Let us clarify the reason. After the BH formation, there is a"settlement" phase due to "gravitational recoil". As one is considering SBHs, which arise from a perfect radial gravitational collapse having spherical symmetry, there is no gravitational wave (GW) emission in such a "settlement" phase. In fact, in general relativity GWs are quadrupole waves and a perfect spherical symmetric source cannot emit GWs [7]. On the other hand, from the quantum point of view, one can consider the BH formation as subsequent absorptions of separated particles starting from a BH having null mass, where one recalls that the statement "a BH having null mass" means the situation of the gravitational collapse before the formation of the first event horizon. In that case, the approach in [12, 53] implies that absorption of particles by the BH generates perturbations and that the BH back reaction to such perturbations is given by QNMs. In other words, the "settlement" phase after the BH formation is completely described by the BH QNMs "triggered" by the absorption of particles. Now, let us ask: when will the "settlement" phase end? In order to answer this question one has to recall that a BH is not immersed in vacuum. Instead, it is immersed in the thermal bathroom of the Cosmic Background Radiation (CBR). Thus, it will continue to absorb CBR photons (and other potential external particles) until the CBR temperature will be higher than the BH Hawking temperature. In other words, the "settlement" phase will end when the BH Hawking temperature will become higher than the CBR temperature. But, at that point, the BH will start to emit Hawking radiation and the BH QNMs will be "triggered" by the absorption of Hawking quanta having negative energies. The process will continue again and again, until the Planck distance and the Planck mass are approached. At that point, the generalized uncertainty principle (GUP) prevents the BH from its total evaporation by stopping Hawking's evaporation process in exactly the same way that the usual HUP prevents the hydrogen atom from total collapse [72]. Thus, the BH arrives at a minimum energy level which is compatible with the GUP. Such an energy level will be rigorously obtained in next Sections where it will be also shown that it consists of a negative energy having the absolute value of the Planck energy. Thus, from the above discussion one argues that the *sole* BH at rest is the BH having null mass and and one must consider absorptions instead of emissions. This means that one must replace $M \to 0$ and $-\omega_n \to M$ in the right hand side of Eq. (8). One also recalls that, differently from the works [27–42] the principal quantum number increases for absorptions instead of emissions. Thus, one obtains

$$\omega_n = \frac{in}{2M} = \frac{in}{4M_E} \tag{14}$$

and this means that, again, one gets $M_E = \frac{M}{2}$ because one is interested in the energy eigenvalues as being absorptions starting from the BH formation. One can still semi-classically interpret the quasi-normal frequency (14) as being a particle, the "electron", quantized on a circle. On the other hand, it will be shown in the following that, in order to get consistence with the exact quantum solution of BH energy levels, which will be derived in next Section, one assigns to the QNMs in Eq. (14) the half of the total effective energy of the two-particle quantum system, obtaining [48]

$$\omega_n = \frac{E_n}{2} = \frac{in}{2M} = \frac{in}{4M_E}. \tag{15}$$

In next Section it will be also shown that, differently from previous literature on BH quantum physics, as the BH is a bound quantum system, the total BH energy is given by $E = -\frac{M_E}{2}$, being negative and different from the BH inert mass M [48]. Thus,one gets from Eq. (15)

$$\frac{M_E}{4} = \left|\frac{in}{4M_E}\right|, \tag{16}$$

which is easily solved obtaining

$$M_n = 2\sqrt{n}. \tag{17}$$

Remarkably, this mass spectrum is consistent with the BH mass spectrum found by Bekenstein in 1974 [9]. Bekenstein indeed obtained $M_n = \sqrt{\frac{n}{2}}$ by using the Bohr-Sommerfeld quantization condition because he argued that the SBH behaves as an adiabatic invariant. Thus, in an intriguing way, one interprets Eq. (15) in terms of a particle, the "electron", quantized on a circle of length

$$L = \frac{1}{T_E} = 8\pi M_E = 8\pi\sqrt{n}, \tag{18}$$

which represents the "electron" travelling in circular orbits around the "hydrogen nucleus". This semi-classical physical interpretation of the QNMs in Eq. (14) is of a collection of damped harmonic degrees of freedom, which releases an intuitive physical picture of a BH as a whole [53]. The larger is $|\omega_n|$, the shorter the lifetime, as one expects from physical intuition [53]. Thus, the increasing of absorptions "triggers" shorter and shorter lived modes, while subsequent emissions of Hawking quanta "trigger" longer and longer lived modes.

3. Schrodinger Theory of the Schwarzschild Black Hole

The Schrodinger theory of the SBH has been recently developed by the Author and Collaborators in the recent works [45–48]. Thus, in this Section we closely follow such papers.

3.1. Quantization of the Historical Oppenheimer and Snyder Gravitational Collapse via Feynman's Path Integral Approach

In 1993, famous Einstein's collaborator Nathan Rosen derived an intuitive and intriguing, but also rigorous, approach to the quantization of closed homogeneous and isotropic cosmological models [49]. His idea was simple but very clever. He multiplied both of the sides of the dynamical Einstein-Friedmann equation (3) in [49] by the quantity $\frac{1}{2}mc^2$, where m is the mass of the Universe, and he considered the new obtained equation (8) in [49] like the energy equation for a particle in one-dimensional motion. Such an idea permitted Rosen to find that in the case of a Universe filled with pressureless matter (dust) the equation is like that for the s states of a hydrogen-like atom [49]. Rosen's approach has been applied to the historical Oppenheimer and Snyder gravitational collapse (the simple case of a pressureless "*star of dust*") in [45, 46]. The classical framework of this kind of gravitational collapse is well known [7, 45, 49, 50, 73]. From the historical point of view, it was originally analysed in the famous paper of Oppenheimer and Snyder [50]. A different approach has been instead developed by Beckerdoff and Misner [73]. More recently, a non-linear electrodynamics Lagrangian has been added in this collapse's framework by the Author and a collaborator in [74]. This different approach permitted a way to remove the BH singularity at the classical level [74]. For the interior of the collapsing star, one uses the well-known Friedmann-Lemaitre-Robertson-Walker (FLRW) line-element with comoving hyper-spherical coordinates χ, θ,

φ [7]. Thus, one writes down [7]

$$ds^2 = d\tau^2 + a(\tau)(-d\chi^2 - \sin^2\chi(d\theta^2 + \sin^2\theta d\varphi^2), \tag{19}$$

where the origin of coordinates is set at the centre of the star, and $a(\tau)$ is the scale factor given by the familiar cycloidal relation [7]

$$\begin{aligned} a &= \tfrac{1}{2}a_m\,(1+\cos\eta)\,, \\ \tau &= \tfrac{1}{2}a_m\,(\eta+\sin\eta) \end{aligned} \tag{20}$$

The density is given by [7]

$$\rho = \left(\frac{3a_m}{8\pi}\right)a^{-3} = \left(\frac{3}{8\pi a_m^2}\right)\left[\frac{1}{2}\,(1+\cos\eta)\right]^{-3}. \tag{21}$$

Setting $\sin^2\chi$ one chooses the case of positive curvature, which corresponds to a gas sphere whose dynamics begins at rest with a finite radius, and, in turn, it is the only one of interest [7]. Thus, the choice $k = 1$ is made for dynamical reasons (the initial rate of change of density is null, that means "momentum of maximum expansion" [7]), but the dynamics also depends on the field equations. In order to discuss the simplest model of a *"star of dust"*, that is, the case of zero pressure, one sets the stress-energy tensor as [7]

$$T = \rho u \otimes u, \tag{22}$$

where ρ is the density of the collapsing star and u the four-vector velocity of the matter. On the other hand, the external geometry is given by the Schwarzschild line-element [7]

$$ds^2 = \left(1 - \frac{2M}{r}\right)dt^2 - r^2\left(\sin^2\theta d\varphi^2 + d\theta^2\right) - \frac{dr^2}{1-\frac{2M}{r}}, \tag{23}$$

where M is the total mass of the collapsing star. As there are no pressure gradients, which can deflect the the motion of the particles, the particles on the surface of any ball of dust move along radial geodesics in the exterior Schwarzschild space-time [7]. Considering a ball which begins at rest with finite radius (in terms of the Schwarzschild radial coordinate) $r = r_i$ at the

(Schwarzschild) time $t = 0$, the geodesics motion of its surface is given by the following equations [7]:

$$r = \frac{1}{2} r_i \left(1 + \cos \eta\right), \tag{24}$$

$$\begin{aligned} t = 2M \ln \left[\frac{\sqrt{\frac{r_i}{2M} - 1} + \tan\left(\frac{\eta}{2}\right)}{\sqrt{\frac{r_i}{2M} - 1} - \tan\left(\frac{\eta}{2}\right)} \right] \\ + 2M \sqrt{\frac{r_i}{2M} - 1} \left[\eta \quad \left(\frac{r_i}{4M}\right) \left(\eta + \sin \eta\right) \right]. \end{aligned} \tag{25}$$

The proper time measured by a clock put on the surface of the collapsing star is [7]

$$\tau = \sqrt{\frac{r_i^3}{8M}} \left(\eta \qquad \eta\right). \tag{26}$$

The collapse begins for $r = r_i$, $\eta = \tau = t = 0$, and terminates at the singularity $r = 0$, $\eta = \pi$ after a duration of proper time measured by the falling particles [7]

$$\Delta \tau = \pi \sqrt{\frac{r_i^3}{8M}}, \tag{27}$$

which coincidentally corresponds, as it is well known, to the interval of Newtonian time for free-fall collapse in Newtonian theory. Differently from the cosmological case, where the solution is homogeneous and isotropic everywhere, here the internal homogeneity and isotropy of the FLRW line-element are broken at the star's surface, that is, a some radius $\chi = \chi_0$ [7]. At that surface, which is a 3-dimensional world tube enclosing the star's fluid [7], the interior FLRW geometry must match smoothly the exterior Schwarzschild geometry [7]. One considers a range of χ given by $0 \leq \chi \leq \chi_0$, with $\chi_0 < \frac{\pi}{2}$ during the collapse [7]. For the pressureless case the match is possible [7]. The external Schwarzschild solution predicts indeed a cycloidal relation for the star's circumference [7]

$$\begin{aligned} C = 2\pi r = 2\pi \left[\tfrac{1}{2} r_i \left(1 + \cos \eta\right)\right], \\ \tau = \sqrt{\tfrac{r_i^3}{8M}} \left(\eta \qquad \eta\right). \end{aligned} \tag{28}$$

The interior FLRW predicts a similar cycloidal relation [7]

$$\begin{aligned} C = 2\pi r = 2\pi a \sin\chi_0 = \pi \sin\chi_0 a_m \left(1 + \cos\eta\right), \\ \tau = \tfrac{1}{2} a_m \left(\eta + \sin\eta\right). \end{aligned} \tag{29}$$

Therefore, the two predictions agree perfectly for all time if and only if [7]

$$\begin{aligned} r_i = a_0 \sin\chi_0, \\ M = \tfrac{1}{2} a_0 \sin^3\chi_{0,} \end{aligned} \tag{30}$$

where r_i and a_0 are the values of the Schwarzschild radial coordinate in Eq. (23) and of the scale factor in Eq. (19) at the beginning of the collapse, respectively. Thus, Eqs. (30) represent the requested match, while the Schwarzschild radial coordinate, in the case of the matching between the internal and external geometries, is [7]

$$r = a \sin\chi_0. \tag{31}$$

One notes that the initial conditions on the matching are the simplest possible that could be relaxed, still having a continuous matching without extra surface terms. In fact, taking the interior solution to be homogeneous requires very fine tuned initial conditions for the collapse and the dynamics of the edge. Thus, on one hand, further analyses for a better characterization of the initial conditions on the matching between the internal and external geometries could be the object of future works. On the other hand, despite the analysis of this Book Chapter is not the most general possible, one recalls that, as it has been previously stressed in the Introduction of this Book Chapter, the problem of the BH quantization is one of the most important of modern theoretical physics. Thus, in order to solve such a fundamental problem, one must start from the simplest case rather than from more complicated ones. This is in complete analogy with the history of general relativity. The first solution of Einstein field equations was indeed the Schwartzschild solution, but it was not a general, rotating solution which included cosmological term or other sources of dark energy, as well the corresponding gravitational collapse developed by Oppenheimer and Snyder [50] did not include a class of non-homogeneous models. Thus, in this novel approach to the BH quantization, here one starts from the simplest conditions rather than from more complicated ones. The initial conditions on the matching that are applied here are exactly the ones proposed by Oppenheimer and Snyder in their historical paper on the gravitational collapse. It is well known that the final result of the gravitational collapse studied by Oppenheimer and Snyder is the Schwarzschild BH [7]. Hence, one also sees what happens when the star is completely collapsed, i.e. when the star is a BH. By inserting $r_i = 2M = r_g$, where r_g is the gravitational radius (the Schwarzschild radius), in Eqs. (30), one gets $\sin^2\chi_0 = 1$. Thus, as the range $\chi > \frac{\pi}{2}$ must be discarded [7], one concludes that it is $\chi_0 = \frac{\pi}{2}$ for a BH. The quantization approach derived by Rosen in [49] has been applied to the above case in [45, 46]. Here it will be shown that the same result can be obtained via Feynman's path integral approach. One starts by rewriting the FLRW line-element (19) in spherical coordinates and comoving time as [45, 49]

$$ds^2 = d\tau^2 - a^2(\tau)\left(\frac{dr^2}{1-r^2} + r^2 d\theta^2 + r^2\sin^2\theta d\varphi^2\right). \tag{32}$$

The Einstein field equation [45, 49]

$$G_{\mu\nu} = -8\pi T_{\mu\nu} \tag{33}$$

gives the relations (one assumes zero pressure) [45, 49]

$$\begin{aligned}\dot{a}^2 &= \tfrac{8}{3}\pi a^2\rho - 1\\ \ddot{a} &= -\tfrac{4}{3}\pi a\rho\end{aligned} \tag{34}$$

with $\dot{a} = \frac{da}{d\tau}$. For consistency, one gets [45, 49]

$$\frac{d\rho}{da} = -\frac{3\rho}{a}, \tag{35}$$

which, when integrated gives [45, 49]

$$\rho = \frac{C}{a^3}. \tag{36}$$

In the collapse case, C is determined by the initial conditions as [7]

$$C = \frac{3a_0}{8\pi}. \tag{37}$$

Now, one rewrites Eq. (36) as

$$\rho = \frac{3a_0}{8\pi a^3}. \tag{38}$$

By multiplying the first of (34) for $M/2$ one gets [45, 47]

$$\frac{M\dot{a}^2}{2} - \frac{4}{3}\pi M a^2 \rho = -\frac{M}{2}, \tag{39}$$

One considers the standard Einstein - Hilbert Lagrangian [2]

$$L_{EH} = \frac{\sqrt{-g}R}{16\pi}. \tag{40}$$

By using the FLRW line-element (32) one gets

$$L_{FLRW} = \dot{a}^2 + \frac{8}{3}\pi a^2 \rho. \tag{41}$$

In order to have consistence with Feynman's PhD thesis [75], one rescales the Lagrangian (41) as

$$L = \frac{M\dot{a}^2}{2} + \frac{4}{3}M\pi a^2 \rho. \tag{42}$$

In this way, the Lagrangian is rewritten in Feynman's form [75]

$$L = \frac{M\dot{a}^2}{2} - V(a), \tag{43}$$

where

$$V(a) \equiv -\frac{4}{3}M\pi a^2 \rho. \tag{44}$$

Inserting Eq. (38) into Eq. (44) one obtains

$$V(a) = -\frac{M a_0}{2a}. \tag{45}$$

The energy function associated to the Lagrangian is

$$E = \frac{\partial L}{\partial \dot{a}}\dot{a} - L. \tag{46}$$

Then, by inserting Eq. (42) in Eq. (46) and by using the first of Eqs. (34) one gets

$$E = -\frac{M}{2}. \tag{47}$$

By taking an infinitesimal time interval τ to $\tau + \delta\tau$, the wave function Ψ for the system at the time $\tau + \delta\tau$ in terms of its value at time τ is determined by the integral equation [75]

$$\Psi(Q, \tau + \delta\tau) = \int \exp\left[i\delta\tau L\left(\frac{Q-q}{\delta\tau}, Q\right)\right] \Psi(q,\tau) \frac{\sqrt{g(q)}dq}{A(\delta\tau)}, \tag{48}$$

where, the transformation function $\left(q'_{\tau+\delta\tau} \equiv Q | q'_\tau \equiv q\right)$ which connects the representations referring to the two different times τ and $\tau + \delta\tau$, corresponds in the classical theory to $\exp(i\delta\tau L)$. $\sqrt{g(q)}dq$ is the volume element in q-space and $A(\delta\tau)$ is a suitable normalization constant see [75] for details. In accordance with Eq. (48), then the wave function for the system must satisfy (where one writes ε for $\delta\tau$) for infinitesimal ε, the equation [75]

$$\Psi(a, \tau + \varepsilon) = \int \exp i\varepsilon \left[\frac{M}{2}\left(\frac{a-y}{\varepsilon}\right)^2 - V(a)\right] \Psi(y,\tau) \frac{dy}{A}. \tag{49}$$

Now, one replaces $y = \eta + a$ in the integral. Thus, one writes [75]

$$\Psi(a, \tau + \varepsilon) = \int \exp i \left[\frac{M}{2}\frac{\eta^2}{\varepsilon} - \varepsilon V(a)\right] \Psi(\eta + a, \tau) \frac{d\eta}{A}. \tag{50}$$

Only values of η close to zero will contribute to the integral [75]. In fact, for small ε, other values of η make the exponential oscillate so rapidly that there will be little contribution to the integral [75]. Then, one expands $\Psi(\eta + a, \tau)$ in a Taylor series around $\eta = 0$ [75]. After rearranging, one gets the integral [75]

$$\begin{gathered} \Psi(a, \tau + \varepsilon) = \frac{\exp -(i\varepsilon V(a))}{A} \int \left[\exp\left(i\frac{M}{2}\frac{\eta^2}{\varepsilon}\right)\right] * \\ \left[\Psi(a,\tau) + \eta\frac{\partial\Psi(a,\tau)}{\partial a} + \frac{\eta^2}{2}\frac{\partial^2\Psi(a,\tau)}{\partial a^2} + ...\right] d\eta. \end{gathered} \tag{51}$$

Now, from Pierces integral tables in Feynman's PhD thesis [75], one gets

$$\int_{-\infty}^{\infty} \exp\left(i\frac{M}{2}\frac{\eta^2}{\varepsilon}\right) d\eta = \sqrt{\frac{2\pi i\varepsilon}{M}}, \tag{52}$$

and, by differentiating both sides with respect to M, one finds [75]

$$\int_{-\infty}^{\infty} \exp\left(i\frac{M}{2}\frac{\eta^2}{\varepsilon}\right)\eta^2 d\eta = \sqrt{\frac{2\pi i\varepsilon}{M}}\frac{\varepsilon i}{M}. \tag{53}$$

The integral with η in the integrand is the integral of an odd function. Then, it is zero [75]. Hence,

$$\Psi\left(a,\tau+\varepsilon\right) = \frac{\sqrt{\frac{2\pi i\varepsilon}{M}}}{A}\int_{-\infty}^{\infty} \exp-\left[i\varepsilon V(a)\right]\left[\Psi\left(a,\tau\right) + \frac{\varepsilon i}{2M}\frac{\partial^2\Psi\left(a,\tau\right)}{\partial a^2} + ...\right]. \tag{54}$$

The left hand side of Eq. (54) approaches $\Psi\left(a,\tau\right)$ for very small ε. Thus, if one wants the equality to hold one must choose [75]

$$A\left(\varepsilon\right) = \sqrt{\frac{2\pi i\varepsilon}{M}}. \tag{55}$$

If one expands both sides Eq. (54) in powers of ε up to the first order, one finds [75]

$$\Psi\left(a,\tau\right) + \varepsilon\frac{\partial\Psi\left(a,\tau\right)}{\partial t} = \Psi\left(a,\tau\right) - i\varepsilon V(a)\Psi\left(a,\tau\right)\frac{\varepsilon i}{2M}\frac{\partial^2\Psi\left(a,\tau\right)}{\partial a^2}, \tag{56}$$

and therefore [75],

$$i\frac{\partial\Psi}{\partial\tau} = -\frac{1}{2M}\frac{\partial^2\Psi}{\partial a^2} + V\Psi. \tag{57}$$

In [45] Eq. (57) has been obtained via Rosen's quantization approach [49]. In this Book Chapter, it has been obtained via Feynman's path integrals approach. One can state that, perhaps, Rosen's approach is more intuitive, while Feynman's approach is more rigorous from the mathematical point of view. But what is really important from the physical point of view is that both of the approaches generate the same results for the quantization of the Oppenheimer and Snyder gravitational collapse.

3.2. Quantum Black Holes

Now, one can closely follow the analysis in [45] in order to get the SBH Schrodinger equation from Eq. (57). One considers the case of a completely collapsed star, i.e. a BH, which means $\chi_0 = \frac{\pi}{2}$, $r = a$ and $r_i = a_0 = 2M = r_g$, in Eqs. (30), see the discussion below Eq. (31). Hence, setting

$$\Psi = rX, \tag{58}$$

Eq. (57) becomes

$$-\frac{1}{2M}\left(\frac{\partial^2 X}{\partial r^2} + \frac{2}{r}\frac{\partial X}{\partial r}\right) + VX = EX, \tag{59}$$

with the associated potential

$$V(r) = -\frac{M^2}{r}. \tag{60}$$

Equations (60) and (59) should be the exact gravitational potential energy and Schrodinger equation of the Schwarzschild BH interpreted as "gravitational hydrogen atom", respectively. Actually, a further final correction is needed. To clarify this point, let us compare Eq. (60) with the analogous potential energy of an hydrogen atom which is [76]

$$V(r) = -\frac{e^2}{r}. \tag{61}$$

Equations (60) and (61) are formally identical, but there is an important difference. In the case of Eq. (61) the electron's charge is constant for all the energy levels of the hydrogen atom. Instead, in the case of Eq. (60), based on the emissions of Hawking quanta or on the absorptions of external particles, the BH mass changes during the jumps from an energy level to another. In fact, such a BH mass decreases for emissions and increases for absorptions. Thus, one must also consider this dynamical behavior of quantum BHs. A good way to take into account this dynamical behavior is by using the *BH effective state* described in Section 2. Although the analysis in Section 2 has been realized for emissions of particles, one immediately argues for symmetry that the same analysis works also in the case of absorptions of external particles, which can be considered as emissions having opposite signs [45–48]. Thus, the effective quantities (6) become

$$M_E \equiv M + \frac{\omega}{2}, \; r_E \equiv 2M_E, \tag{62}$$

and now they represents the BH mass and the BH horizon *during* the BH expansion, i.e., *during* the absorption of the particle. Via Hawking's periodicy argument [60] one argues that, in order to take the BH dynamical behavior into due account, one must replace the BH mass M with the BH effective mass M_E in Eqs. (60) and (59), obtaining [45]

$$V(r) = -\frac{M_E^2}{r}, \tag{63}$$

$$-\frac{1}{2M_E}\left(\frac{\partial^2 X}{\partial r^2} + \frac{2}{r}\frac{\partial X}{\partial r}\right) + VX = EX, \tag{64}$$

with the connection between the energy eigenvalues and the SBH effective mass [45]

$$E_n = -\frac{r_E^2 M_E^3}{8n^2}, \tag{65}$$

and the SBH total energy [45]

$$E = -\frac{M_E}{2}. \tag{66}$$

Now, from the quantum point of view, one wants to obtain the energy eigenvalues as being absorptions starting from the BH formation, that is from the BH having null mass. This implies that we must replace $M \to 0$ and $\omega \to M$ in Eq. (62) [45–48]. Thus, one gets [45–48]

$$M_E \equiv \frac{M}{2},\ r_E \equiv 2M_E = M. \tag{67}$$

One inserts Eqs. (66) and (67) into Eq. (65), obtaining the BH mass spectrum as

$$M_n = 2\sqrt{n}, \tag{68}$$

and by using Eq. (66) one finds the BH energy levels as

$$E_n = -\frac{1}{2}\sqrt{n}. \tag{69}$$

Eq. (68) is exactly Eq. (17) of Section 2. Thus, the full quantum analysis in this Section is completely consistent with the semiclassical approximation (Bohr-like approach) in Section 2. One notes again that the result of Eq. (68) is consistent also with Bekenstein's original result in [9]. A quantum description of BH in terms of quantum membranes has been conjectured by Maggiore [26], who obtained the energy spectrum

$$E_n = \sqrt{\frac{A_0 n}{16\pi}}. \tag{70}$$

Thus, he was forced to set $A_0 = 8\pi$ in order to find Bekenstein's result in [9]. One sees that, in its absolute value, the result of Eq. (69) is consistent with Maggiore's result by setting $A_0 = 16\pi$. On the other hand, both Bekenstein and Maggiore used heuristic analysis, approximations and/or conjectures. Instead, Eq. (68) has been obtained through an exact quantization process. In addition, neither Bekenstein nor Maggiore realized that the BH energy spectrum must have negative eigenvalues because the "gravitational hydrogen atom" is a quantum system composed by bound states. The mass spectrum of Eq. (68) is also consistent with other BH mass spectra in the literature, see for example [13–16]. After Bekenstein and Maggiore also the Authors of previous literature did not realize that the BH energy spectrum *must have negative eigenvalues*. This is one of the most important results in [45–48] that we review in this Book Chapter. The total BH energy $E = -\frac{M}{4}$ *is negative and different* from the BH inert mass M. In fact, the above quantization procedure splits the classical BH in a two-particle quantum system having a gravitational negative energy given by Eq. (63). It is indeed well known that the bound states of physical systems must have negative energies.

Let us discuss in a deeper way the Schrodinger equation (65) and its solution. One considers the analogy between the potential energy of an hydrogen atom, given by Eq. (61) and the effective potential energy of the "gravitational hydrogen atom" given by Eq. (63). Equation (61) represents the interaction between the nucleus of the hydrogen atom, having a charge e and the electron, having a charge $-e$. Equation (63) represents the interaction between the nucleus of the "gravitational hydrogen atom", i.e. the BH, having an effective, dynamical mass M_E, and another, mysterious, particle, i.e. the "electron" of the "gravitational hydrogen atom" having again an effective, dynamical mass M_E. Thus, let us ask: what is the "electron" of the BH? An intriguing answer to this question has been given in Section 2 of this paper via the semiclassical Bohr-like approximation. For large values of the principal quantum number n, the BH QNMs, "triggered" by emissions (Hawking radiation) and absorption of external particles, represent the "electron" which jumps from a level to another one, and the absolute values of the QNMs frequencies, represent the energy "shells" of the "gravitational hydrogen atom". Hence, the QNM jumping from a level to another one has been interpreted in Section 2 in terms of a particle quantized on a circle, which is analogous to the electron travelling in circular orbits around the hydrogen nucleus, similar in structure to the solar system, of Bohr's semiclassical model of the hydrogen atom [43, 44]. The analogy between the quantum SBH and the hydrogen atom due to the similarity of their respective potential energies suggests to extend the interpretation of the "electron" in terms of the quantum oscillations of the BH horizon also to the full quantum analysis of this section. One can indeed consider the de Broglie hypothesis [51] and the wave nature of the BH "electron". This means that such a particle does not orbit the nucleus in the same way as a planet orbits the Sun, but instead exists as a standing wave. The correct analogy is that of a large and often oddly shaped "atmosphere" (the BH "electron"), distributed around a relatively tiny planet (the BH "nucleus"). The correct physical interpretation of such an "atmosphere" is nothing else than the BH horizon modes.

Another fundamental issue is the following. The quantum BH expressed by the system of equations from Eq. (63) to Eq. (66) seems to be *non-singular*. It is indeed well known that, in the classical general relativistic framework, in the internal geometry all time-like radial geodesics of the collapsing star terminate after a lapse of finite proper time in the termination point $r = 0$ and it is impossible to extend the internal space-time manifold beyond that termination point [7]. Thus, the point $r = 0$ represents a singularity based on the rigorous definition by Schmidt [77]. But what happens in the quantum framework analysed in this Book Chapter is completely different. The above analysis has shown that the classical SBH has been split in a two-particle system where the two components strongly interact with each other through a quantum gravitational interaction. The system that has been analysed is indeed formally equal to the well known system of two quantum particles having finite distance with the mutual attraction of the form $1/r$ [76]. Such two particles are the "nucleus" and the "electron" of the "gravitational hydrogen atom". What is the meaning of the word "particle" in a quantum framework? Quantum particles remain in an uncertain, non-deterministic, smeared, probabilistic wave-particle orbital state [76]. Then, they cannot be localized in a particular "termination point where all time-like radial geodesics terminate" [7]. Such a localization is also in contrast with the Heisenberg uncertainty principle (HUP). The HUP says indeed that either the location or the momentum of a quantum particle such as the BH "electron" can be known as precisely as desired, but as one of these quantities is specified more precisely, the value of the other becomes increasingly indeterminate [45]. This is not simply a matter of observational difficulty, but rather a fundamental property of nature. What this means is that within the tiny confines of the “gravitational atom”, the “electron” cannot really be regarded as a "point-like particle" having a definite energy and location [45]. Thus, it is somewhat misleading to talk about the BH “electron” “falling into” the BH “nucleus”. In other words, the Schwarzschild radial coordinate in the system of equations from Eq. (63) to Eq. (66) CANNOT become equal to zero. The GUP makes this last statement even stronger. In fact, it can be written down in the general form [45]

$$\triangle\, x \triangle\, p \geq \frac{1}{2}\left[1+\eta\,(\triangle\, p)^{2}+...\right]. \tag{71}$$

The GUP expressed by Eq. (71) implies a non-zero lower bound on the minimum value of the uncertainty on the particle’s position $(\triangle\, x)$ which is of order of the Planck length [78] . In other words, the GUP implies the existence of a minimal length in quantum gravity. One also notes another important difference between the hydrogen atom of quantum mechanics [76] and the “gravitational hydrogen atom” in the present analysis. In the standard hydrogen atom the nucleus and the electron are different particles. In the quantum BH here analysed they are equal particles instead, as one easily checks in the system of equations from Eq. (63) to Eq. (66). Thus, the “nucleus” and the “electron” can be mutually exchanged without varying the physical properties of the system. Hence, the quantum state of the two particles is even more uncertain, more non-deterministic, more smeared and more probabilistic than the corresponding quantum states of the particles of the hydrogen atom [45]. These quantum argumentations seem to be strong arguments in favour of the non-singular behavior of the SBH in a quantum framework. The results obtained here are also in agreement with the general conviction that quantum gravity effects become fundamental in the presence of strong gravitational fields. In a certain sense, the results in [45–48], that have been reanalysed in this Section, permit to “see” into the SBH. The above analysis can be extended to the Reissner-Nordstrom BH, see [46] for details.

Another intriguing issue, originally developed by the Author in [47], is the following. On one hand, Eq. (64) seems simpler than the corresponding Schrodinger equation for the hydrogen atom because there is no angular dependence; on the other hand it seems also more complicated, because, differently from the Schrodinger equation for the hydrogen atom, where the electron's charge is constant, in Eq. (64) the BH mass variates, and this is exactly the reason because the BH effective mass, that is indeed a dynamical quantity, has been introduced. One also recalls that $e^2 = \alpha$ is the fine structure constant, which, in standard units, combines the constants $\frac{e^2}{4\pi\epsilon_0}$ from electromagnetism, $\hbar$ from quantum mechanics, and the speed of light c from relativity, into the dimensionless irrational number $\alpha \simeq \frac{1}{137{,}036}$, which is one of the most important numbers in Nature. Thus, from the present approach one argues that the gravitational analogous of the fine structure constant is not a constant. It is a dynamical quantity instead. In fact, if one labels the "gravitational fine structure constant" as α_G, by confronting Eqs. (63) and (61) one gets [47]

$$\alpha_G = \left(\frac{M_E}{m_p}\right)^2, \tag{72}$$

in standard units, where m_p is the Planck mass. It will be indeed shown that α_G has a spectrum of values which coincides with the set of natural numbers $\mathbb{N}$. In analogous way, the Rydberg constant, $R_\infty = \frac{1}{2}m_e\alpha^2$, is defined in terms of the electron mass m_e and on the fine structure constant. In the current gravitational approach the effective mass and the "charge" are the same. Hence, the quantum gravitational analogous of the Rydberg constant is [47]

$$(R_\infty)_G = \frac{M_E^5}{2m_p^5 l_p}. \tag{73}$$

in standard units, where l_p is the Planck length. Again, it will be shown that $(R_\infty)_G$ is not a constant but a dynamical quantity having a particular spectrum of values.

By introducing the variable y [47]

$$y(r) \equiv rX \tag{74}$$

Eq. (64) becomes [47]

$$-\left(\frac{1}{2M_E}\frac{d^2}{dr^2} + \frac{M_E^2}{r} + E\right)y = 0. \tag{75}$$

Setting $y' \equiv \frac{d}{dr}$ and using $E = -\frac{M_E}{2}$, Eq. (75) can be rewritten as [47]

$$y'' + M_E^2\left(\frac{2M_E}{r} - 1\right)y = 0. \tag{76}$$

As it is $E < 0$, the asymptotic form of the solution, which is regular at the origin, is a linear combination of exponentials $\exp(M_E r)$, $\exp(-M_E r)$. If one wants this solution to be an acceptable eigensolution, the coefficient in front of $\exp(M_E r)$ must vanish. This happens only for certains discrete values of E. Such values will be the energies of the discrete spectrum of the SBH and the corresponding wave function represents one of the possible SBH bound states. If one makes the change of variable [47]

$$x = 2M_E r \tag{77}$$

Eq. (76) results equivalent to [47]

$$\left[\frac{d^2}{dx^2} + \frac{M_E^2}{x} - \frac{1}{4}\right] y = 0, \tag{78}$$

and y is the solution which goes as x at the origin. For x very large, it increases exponentially, except for certain particular values of the SBH effective mass where it behaves as $\exp\left(-\frac{x}{2}\right)$. One wants to determine such special values and their corresponding eigenfunctions. One starts to perform the change of function [47]

$$y = x \exp\left(-\frac{x}{2}\right) z(x), \tag{79}$$

which changes Eq. (78) to [47]

$$\left[\frac{d^2}{dx^2} + (2 - x)\frac{d}{dx} - \left(1 - M_E^2\right)\right] z = 0. \tag{80}$$

This last equation is a Laplace-like equation. Within a constant, one finds only a solution which is positive at the origin. All the other solutions have a singularity in x^{-1}. One can show that this solution is the confluent hypergeometric series [47]

$$Z = \sum_{i=1}^{\infty} \frac{\Gamma\left(1 + i - M_E^2\right)}{\Gamma\left(1 - M_E^2\right)} \frac{1}{(1+i)!} \frac{x^i}{i!}. \tag{81}$$

In fact, one expands the solution of Eq. (80) in Mc Laurin series at the origin as [47]

$$z = 1 + \alpha_1 x + \alpha_2 x^2 + ... \alpha_i x^i + ... \tag{82}$$

By inserting Eq. (82) in Eq. (80) one writes the LHS in terms of a power series of x. All the coefficients of this expansion must be null. Thus [47],

$$\begin{aligned} 2\alpha_1 &= 1 - M_E^2 \\ 2 * 3\alpha_2 &= \left(2 - M_E^2\right)\alpha_1 \\ &.... \\ i\left(1+i\right)\alpha_i &= \left(i - M_E^2\right)\alpha_{i-1}. \end{aligned} \tag{83}$$

Then, one gets [47]

$$\alpha_i = \frac{\left(i - M_E^2\right)}{(1+i)} \frac{\left(i - 1 - M_E^2\right)}{(1+i-1)} \cdots \frac{1 - M_E^2}{1+1} \frac{1}{i!}, \tag{84}$$

which is exactly the coefficient of x^i in the expansion (82).

From a mathematical point of view the series (81) is infinite and behaves as [47]

$$\frac{\exp x}{x^{\left(1+M_E^2\right)}}$$

for large x. Consequently, y behaves in the asymptotic region as [47]

$$\frac{\exp \frac{x}{2}}{x^{M_E^2}}$$

and cannot be, in general, an eigensolution. On the other hand, for particular values of M_E^2 all the coefficients will vanish from a certain order on. In that case, the series (81) reduces to a polynomial. The requested condition is [47]

$$1 - M_E^2 \leq 0, \; with \; \left|1 - M_E^2\right| \in \mathbb{N}, \tag{85}$$

which implies the quantization condition [47]

$$M_E^2 = n = n' + 1, \; with\, n' = 0, 1, 2, 3, ... + \infty. \tag{86}$$

The hypergeometric series becomes a polynomial of degree n' and y behaves in the asymptotic region as [47]

$$\frac{x^n}{\exp \frac{x}{2}}.$$

Then, the regular solution of the SBH Schrodinger equation becomes an acceptable eigensolution. The quantization condition (86) permits to obtain the spectrum of the effective mass as [47]

$$(M_E)_n = \sqrt{n}, \; with\, n = 1, 2, 3, ... + \infty. \tag{87}$$

The quantization condition (86) gives also the spectrum of the "gravitational fine structure constant" that one writes as [47]

$$(\alpha_G)_n = \left[\frac{(M_E)_n}{m_p}\right]^2 = n, \; with\, n = 1, 2, 3, ... + \infty. \tag{88}$$

Thus, the gravitational analogous of the fine structure constant is not a constant. It is a dynamical quantity instead, which has the spectrum of values (88) which coincides with the set of non-zero natural numbers $\mathbb{N} - \{0\}$. In the same way, one finds the spectrum of the "gravitational Rydberg constant" as [47]

$$[(R_\infty)_G]_n = \frac{(M_E)_n^5}{2m_p^5 l_p} = \frac{n^{\frac{5}{2}}}{2}, \; with\, n = 1, 2, 3, ... + \infty. \tag{89}$$

By combining Eqs. (87) and (67) one immediately re-obtains Eq. (68). The (unnormalized) wave function corresponding to each energy level of Eq. (69) is [47]

$$X(r) = \frac{1}{r}\left[\frac{2M_E r}{\exp\frac{x}{2}}\left(\sum_{i=0}^{n-1}(-)^i \frac{(n-1)!}{(n-1-i)!\,(i+1)!i!}\right)(2M_E r)^i\right]. \tag{90}$$

One observes that the number of nodes of the wave function is exactly $n-1$ and that the energy spectrum (69) contains a denumerably infinite number of levels because n can take all the infinite values $n \in \mathbb{N} - \{0\}$. One can calculate the energy jump between two neighboring level as [47]

$$\triangle E = E_{n+1} - E_n = \sqrt{\frac{n}{4}} - \sqrt{\frac{n+1}{4}} = -\frac{1}{4\left(\sqrt{\frac{n}{4}} + \sqrt{\frac{n+1}{4}}\right)}. \tag{91}$$

For large n one obtains [47]

$$\triangle E \simeq -\frac{1}{4\sqrt{n}}$$

when $n \rightarrow +\infty$ one gets $\triangle E \rightarrow 0$. Thus, one finds that the energy levels become more and more closely spaced and their difference tends to $\triangle E = 0$ at the limit, at which point the continuous spectrum of the SBH begins.

One also observes that a two particle Hamiltonian [47]

$$H\left(\vec{p}, r\right) = \frac{p^2}{2M_E} - \frac{M_E^2}{r}, \tag{92}$$

which governs the SBH quantum mechanics, must exist in correspondence of Eqs. (63) and (64). Therefore, the square of the wave function (90) must be interpreted as the probability density of a single particle in a finite volume. Thus, the integral over the entire volume must be normalized to unity as [47]

$$\int dx^3 \left|X\right|^2 = 1. \tag{93}$$

For stable particles, this normalization must remain the same at all times of the SBH evolution. This issue has deep implications for the BH information paradox [57] because it guarantees preservation of quantum information and is consistent with the solution of the paradox which will be discussed in the following. As the wave function (90) obeys the SBH Schrödinger equation (64), this is assured if and only if the Hamiltonian operator (92) is Hermitian [71]. In other words, the Hamiltonian operator (92) must satisfy for arbitrary wave functions X_1 and X_2 the equality [71]

$$\int dx^3 \left[HX_2\right]^* X_1 = \int dx^3 X_2^* H X_1. \tag{94}$$

One notes that both $\vec{p}$ and r are Hermitian operators. Thus, the Hamiltonian (92) will automatically be a Hermitian operator if it is a sum of a kinetic and a potential energy [71]

$$H = T + V. \tag{95}$$

This is always the case for non-relativistic particles in Cartesian coordinates [71] and works also for SBHs.

Thus, it has been shown that the SBH is a well defined quantum system, which obeys Schrodinger's theory.

Now, one recalls that in the SBH the *horizon area* A is related to the mass by the relation $A = 16\pi M^2$. If one assumes that a neighboring particle is captured by the BH causing a transition from the state with n to the state with $n+1$(two neighboring levels) by using the mass spectrum (68), one immediately gets the area quantum as

$$\triangle A_{n\rightarrow n+1}=16\pi\left(M_n-M_{n-1}\right)^2=64\pi\left(n+1-n\right)=64\pi, \qquad (96)$$

which is similar to the original result of Bekenstein [9], but with a different coefficient. This is not surprising, because there is no general consensus on the area quantum, see for example [12–16]. Following [45], for the BH ground state ($n=1$), from Eq. (68) one gets the mass as

$$M_1=2 \qquad (97)$$

in Planck units. Thus, in standard units one gets $M_1=2m_P$, where m_P is the Planck mass, $m_P=2,17645\text{x}10^{-8}Kg$. A total negative energy arising from Eq. (69) and a Schwarzschild radius are associated to the mass (97)

$$E_1=-\frac{1}{2}, \qquad (98)$$

$$r_{g1}=4. \qquad (99)$$

This is the state having minimum mass and minimum energy (the energy of this state is minimum in absolute value; in its real value, being negative, it is maximum). It represents the smallest possible BH. In the case of Bohr's semiclassical model of hydrogen atom, the Bohr radius, which represents the classical radius of the electron at the ground state, is [76]

$$Bohr\ radius=\frac{1}{m_e e^2}, \qquad (100)$$

where m_e is the electron mass. In order to obtain the correspondent "Bohr radius" for the "gravitational hydrogen atom", one needs to replace both m_e and e in Eq. (100) with the effective mass of the BH ground state, which is $\frac{M_1}{2}=1$. Thus, now the "Bohr radius" becomes

$$b_1=1, \qquad (101)$$

which in standard units reads $b_1=l_P$, where $l_P=1,61625\text{x}10^{-35}m$ is the Planck length. Thus, it has been found that the "Bohr radius" associated to the smallest possible BH is equal to the Planck length. The wave-function associated to the BH ground state is [45]

$$\Psi_1=2b_1^{-\frac{3}{2}}r\exp-\left(\frac{r}{b_1}\right)=2r\exp-\left(r\right), \qquad (102)$$

where Ψ_1 is normalized as

$$\int_0^\infty \Psi_1^2 dr = 1. \tag{103}$$

The size of this BH is of the order of

$$\bar{r}_1 = \int_0^\infty \Psi_1^2 r dr = \frac{3}{2} b_1 = \frac{3}{2}. \tag{104}$$

It is also important to recall that in quantum mechanics the Bohr radius (101) gives the radius with the maximum radial probability density instead of its expected radial distance [80]. The latter is indeed 1.5 times the Bohr radius [80]. This depends on the long tail of the radial wave function [80] and it is so given by Eq. (104). Thus, remarkably, Eqs. (101) and (104) are in complete agreement with the standard quantum approach to the hydrogen atom. This cannot be a coincidence. Instead, it is a confirmation of the correctness of the approach in this paper. Hence, an interesting quantum representation of the SBH ground state at the Planck scale has been obtained. This SBH ground state represents the BH minimum energy level which is compatible with the generalized uncertainty principle (GUP) [78]. The GUP indeed prevents a BH from its total evaporation by stopping Hawking's evaporation process in exactly the same way that the usual HUP prevents the hydrogen atom from total collapse [72]. For the sake of completeness, one recalls that today the GUP is used within a large framework of physical processes, see [81–89] and references within for recent developments. One can also easily find the size, that is the expected radial distance, of the quantum SBH excited at the level n as

$$\bar{r}_n = \frac{3}{2}\sqrt{n}. \tag{105}$$

Let us clarify an important point following [48]. It is well known that the Schwarzschild coordinates break down when the radius of the star becomes the gravitational radius, which means that it is no longer possible to match the χ_0 surface with a space-like normal to a constant r surface which is null at the gravitational radius and has a time-like normal for values of r which are less than the gravitational radius. Thus, one stresses here that one never went beyond the gravitational radius. This issue is clarified as follows. The passage from a classical to a quantum analysis splits the classical BH in a two-particle system, the "nucleus" and the "electron". Here the key point is the physical meaning of the second particle, the "electron", during the BH formation and in the following

BH dynamical evolution. From a quantum point of view, the BH formation is interpreted as being the formation of the initial pair composed by the "nucleus" and the "electron". Thus, it represents an absorption from the BH having null mass. This is not an instantaneous process but, instead, it happens in a finite lapse of time. Let us assume that, during a little, but finite, lapse of comoving time (which is the time that we used in the above analysis), say $\triangle\tau$, the gravitational collapse generates a BH having an initial mass $M_n = 2\sqrt{n}$. During $\triangle\tau$ the BH is forming, which means that the two particles are evolving. When the BH is completely formed the two particles are in a stationary state, having an energy $E_n = -\frac{\sqrt{n}}{2}$. It should be noted that in the two-particle quantum system, the two particles are equal and can be mutually exchanged without varying the physical properties of the system. This means that, when the system is in the stationary state, we can assign to the "electron" the half of the total energy of the two-particle quantum system, obtaining

$$\frac{E_n}{2} = -\frac{\sqrt{n}}{4} = -\frac{n}{2M_n}, \tag{106}$$

which corresponds to a particle quantized with an- tiperiodic boundary conditions on a circle of length

$$L = 8\pi M_E = 8\pi\sqrt{n}. \tag{107}$$

Thus, one has full consistency with the semiclassical analysis in Section 2. In fact, Eq. (107) is exactly Eq. (18), and this cannot be a coincidence. This means that the radius of the circle is $r_g = 2M_n$, i.e., it is exactly the gravitational radius. Hence, when the system is in the stationary state, that is, when the BH is completely formed, the distance between the "nucleus" and the "electron" in terms of the Schwarzschild radial coordinate is exactly the gravitational radius. One recalls again the following important point. During both the processes of absorptions of external particles, including the original BH formation, and of emissions of Hawking quanta, the BH horizon is not fixed at a constant distance from the BH "nucleus" [32]. In fact, because of energy conservation, the BH contracts during the emission of a particle and expands during an absorption [32]. Such quantum contractions/expansions are not "one shot processes" [32]. They generate oscillations of the horizon instead [32]. Thus, the "Bohr radius", Eq. (101), and the "expected radial distance", namely Eq. (105), should be interpreted as dynamical quantities characterizing the quantum state of the horizon modes during a transition from a stationary state to another, while the "Bohr orbits" of the stationary states are characterized by Eq. (107). Clearly, if the second particle, that is the "electron", is interpreted in terms of horizon oscillations, then the distance between the two particles is never less than the oscillating gravitational radius.

Let us clarify the results obtained till now. It has been shown that 3 different approaches, that are Rosen's approach in [45, 46], Feynman approach in this Section and the Bohr-like approach in Section 2, give the same results concerning the BH mass and energy spectra. This is quite intriguing and cannot be a coincidence. It also means that one gets 3 different pictures concerning the SBH. In the classical framework of Einstein's general relativity, a BH is a "dead object" with inert mass. It is a definitive prison where anything that enters cannot escape. This means that it can only become more massive and bigger with time. In the semiclassical approximation, Hawking has shown, in his most famous paper [10], that BHs may actually emit quanta. The emission of Hawking radiation implies that BHs lose energy and decrease their mass until eventually evaporating. Hawking's semi-classical picture can be refined through the Bohr-like approach that has been developed in Section 2. The SBH results in some way similar to the Bohr hydrogen atom [43, 44]. The "electron", which is given by the BH QNMs "triggered" by the absorptions of external particles and/or by the emissions of Hawking quanta, travels in circular orbits around the "hydrogen nucleus". The circular orbits become more external in the case of absorptions and more internal in the case of the emissions, see Eq. (18). In other words, the "electron" jumps from an orbit to another one due to absorptions and/or emissions. Finally, a full quantum approach has been developed in [45–48] and in this Section. It has been shown that the SBH is REALLY the gravitational analogous of the hydrogen atom, showing that Bekenstein's original intuition was remarkably correct [11]. The SBH is indeed a well defined quantum, non-singular system, formally equal to the well known system of two quantum particles having finite distance with the mutual attraction of the form $1/r$ [76]. The two particles are really the "nucleus" and the "electron" of the "gravitational hydrogen atom" and they obey Schrodinger theory.

3.3. Restoration of Predictability in Gravitational Collapse

In 1976 S. Hawking claimed that "*Because part of the information about the state of the system is lost down the hole, the final situation is represented by a density matrix rather than a pure quantum state*" (Verbatim from ref. [57]). This was the starting point of the popular "BH information paradox". After Hawking's original claim, enormous time and effort were and are currently devoted to solving the paradox. Consequences of the BH information paradox are indeed not trivial. By assuming that information is lost in BH evaporation, pure quantum states arising from collapsed matter would decay into mixed states arising from BH evaporation [94]. The devastating consequence is that quantum gravity should not be unitary [94]. Various researchers in gravitation worked and still work on this issue. Some of them remain convinced that quantum information is destroyed in BH evaporation. Other ones claim that Hawking's original statement was wrong and information must be, instead, preserved. A popular science book on details of the so-called Black Hole War on the BH information paradox was written by Susskind [95]. In fact, the paradox is today part of physics folklore. Hawking made two famous bets that BH does destroy information. One with Nobel Laureate Kip Thorne like co-signer, with Preskill. The latter with Page, see [95] for details. Readers can find historical notes on the paradox's controversy and on various attempts to solve it in [31, 33, 57], [65–68, 95–97]. Following [48], in this Subsection, the solution which has been originally derived by the Author in [31] will be refined and adapted to the SBH quantum equations that have been obtained in previous analyses in this Book Chapter. Let us consider the instant when the BH Hawking temperature will become higher than the CBR temperature. At that point, the BH will start to emit Hawking radiation. Thus, one assumes a first emission from the BH excited state corresponding to the BH principal quantum number, say m, which has a total energy $E_m = -\frac{1}{2}\sqrt{m}$, given by Eq. (69) and a corresponding total mass $M_m = 2\sqrt{m}$, given by Eq. (68), to a state with $n < m$, which corresponds to a total energy $E_n = -\frac{1}{2}\sqrt{n}$ and to a corresponding total mass $M_n = 2\sqrt{n} < M_m$. From the quantum mechanical point of view, one physically interprets Hawking radiation like energies of quantum jumps among the unperturbed levels of Eq. (69) [31, 48, 97]. In quantum mechanics, time evolution of perturbations can be described by an operator [31, 48, 97]

$$U(t) = \begin{matrix} W(t) \quad for\, 0 \leq t \leq \tau \\ 0 \quad for\, t < 0\, and\, t > \tau. \end{matrix} \tag{108}$$

Then, the complete (time dependent) Hamiltonian is described by the operator [31, 48, 97]

$$H(r,t) \equiv V(r) + U(t), \tag{109}$$

where $V(r)$ is given by Eq. (63). Thus, considering a wave function $\psi(r,t)$, one can write the correspondent *time dependent Schroedinger equation* for the system [31, 48, 97]

$$i\frac{d|\psi(r,t)>}{dt} = [V(r) + U(t)]\,|\psi(r,t)>= H(r,t)|\psi(r,t)>. \tag{110}$$

The state which satisfies Eq. (110) is [9, 56]

$$|\psi(r,t)>= \sum_n a_n(t)\exp(-iE_n t)\,|\varphi_n(r)>, \tag{111}$$

where the $\varphi_n(r)$ are the eigenfunctions of the time independent Schröedinger equation (64) and the E_n are the correspondent eigenvalues. Now, one closely follows [48]. In the basis $|\varphi_n(r)>$, the matrix elements of $W(t)$ can be written as

$$W_{ij}(t) \equiv A_{ij}\delta(t), \tag{112}$$

where $W_{ij}(t) =< \varphi_i(r)|W(t)|\varphi_j(r) >$ and the A_{ij} are real. In order to solve the complete quantum mechanical problem described by the operator (109), one needs to know the probability amplitudes $a_n(t)$ due to the application of the perturbation described by the time dependent operator (108), which represents the perturbation associated to the emission of a Hawking quantum. For $t < 0$, i.e. before the perturbation operator (108) starts to work, the system is in a stationary state $|\varphi_m(t,r)>$, at the quantum level m, with energy $E_m = -\frac{1}{2}\sqrt{m}$, given by Eq. (69). Thus, in Eq. (111) only the term

$$|\psi_m(r,t)>= \exp(-iE_m t)\,|\varphi_m(r)>, \tag{113}$$

is not null for $t < 0$. This implies $a_n(t) = \delta_{nm}$ for $t < 0$. When the perturbation operator (108) stops to work, i.e. after the emission, for $t > \tau$ the probability amplitudes $a_n(t)$ return to be time independent, having the value $a_{m\to n}(\tau)$. In other words, for $t > \tau$ the system is described by the *wave function* $\psi_{final}(r,t)$ which corresponds to the state

$$|\psi_{final}(r,t)>=\sum_{n=1}^{m}a_{m\rightarrow n}(\tau)\exp\left(-iE_{n}t\right)|\varphi_{n}(x)>. \tag{114}$$

Therefore, the probability to find the system in an eigenstate having energy $E_n = -\sqrt{n}$, with $n < m$ for emissions, is given by

$$\Gamma_{m\rightarrow n}(\tau)=|a_{m\rightarrow n}(\tau)|^{2}. \tag{115}$$

By using a standard analysis, one obtains the following differential equation from Eq. (114)

$$i\frac{d}{dt}a_{m\rightarrow n}(t)=\sum_{l=1}^{n}W_{ml}a_{m\rightarrow l}(t)\exp\left[i\left(\Delta E_{l\rightarrow n}\right)t\right]. \tag{116}$$

To first order in $U(t)$, by using the Dyson series [31, 48, 97], one gets the solution

$$a_{m\rightarrow n}=-i\int_{0}^{t}\left\{W_{nm}(t')\exp\left[i\left(\Delta E_{m\rightarrow n}\right)t'\right]\right\}dt'. \tag{117}$$

By inserting Eq. (112) in Eq. (117) one obtains

$$a_{m\rightarrow n}=iA_{nm}\int_{0}^{t}\left\{\delta(t')\exp\left[i\left(\Delta E_{m\rightarrow n}\right)t'\right]\right\}dt'=\frac{i}{2}A_{nm}. \tag{118}$$

Combining this equation with Eq. (144) in [48] and Eq. (115) one gets

$$\alpha\exp\left[16\pi\left(n-m\right)\right].=\tfrac{1}{4}A_{nm}^{2}$$

$$A_{nm}=2\sqrt{\alpha}\exp\left[8\pi\left(n-m\right)\right] \tag{119}$$

$$a_{m\rightarrow n}=-i\sqrt{\alpha}\exp\left[8\pi\left(n-m\right)\right].$$

As $\sqrt{\alpha}\sim 1$, one gets $A_{nm}\sim 10^{-11}$ for $n = m-1$, i.e. when the probability of emission has its maximum value. This implies that second order terms in $U(t)$ are $\sim 10^{-22}$ and can be neglected. Clearly, for $n < m-1$, the approximation is better because the A_{mn} are even smaller than 10^{-11}. Thus, one can write down the final form of the ket representing the state as

$$|\psi_{final}(r,t)>=\sum_{n=1}^{m}-i\sqrt{\alpha}\exp\left[8\pi\left(n-m\right)-iE_{n}t\right]|\varphi_{n}(r)>. \tag{120}$$

The state (120) represents a *pure final state instead of a mixed final state.* Then, the states are written in terms of an *unitary* evolution matrix instead of a density matrix and this implies the fundamental conclusion that information is not lost in BH evaporation. This result is consistent with the idea that Schrödinger equations can be used universally for all dynamics in the universe [98] and dismisses the claim of Hawking [57] that has been cited at the starting of this Section. The final state of Eq. (120) is due to potential arbitrary transitions $m \to n$, with $m > n$. Then, the subsequent *collapse of the wave function* to a new stationary state, at the quantum level n

$$|\psi_n(r,t)>= \exp\left(-iE_n t\right)|\varphi_n(r)>, \tag{121}$$

implies that the wave function of the infalling particle in Hawking's mechanism of particle creation by BH has been transferred to the two-particle system governed by the equations from Eq. (63) to Eq. (66) and it is given by

$$|\psi_{(m\to n)}(r,t)>\equiv \exp\left(-iE_n t\right)|\varphi_n(r)> - \exp\left(-iE_m t\right)|\varphi_m(r)>. \tag{122}$$

This wave function results entangled with the wave function of the particle which propagates towards infinity. Now, it will be shown that this key point solves the entanglement problem connected with the information paradox. This problem of the BH information paradox concerns the entanglement structure of the wave function associated with the particle pair creation [99]. In other terms, in order to solve the paradox, one needs to know the part of the wave function in the interior of the horizon [99], i.e., the part of the wave function associated with the interior, infalling mode. This is exactly the part of the wave function which in the Hawking computation gets entangled with the part of the wave function outside, i.e., the part of the wave function associated to the particle which escapes from the BH [99]. If one ignores this interior part of the wave function, one misses the entanglement completely, and thus fails to understand the paradox [99]. Here the key point is that the particle which falls into the BH transfers its part of the wave function and, in turn, the information encoded in such a part of the wave function, to the two-particle system governed by the equations from Eq. (63) to Eq. (66), that represents the quantum SBH. Hence, the emitted radiation results to be entangled with such a quantum system which concerns the "excited states" of the "gravitational atom". Thus, one argues that the BH response to the absorption of an interior, infalling mode is to add the energy of that interior, infalling mode (and, in turn, its wave function) to the "atom's excited state" corresponding to the energy level E_m, in order to permit it to jump to the "atom's excited state" corresponding to the energy level E_n. In that way, the interior part of the wave function is now "within" the two-particle system governed by the equations from Eq. (63) to Eq. (66), which is now at the quantum level E_n. Let us see this issue in detail. Again, let us consider the instant when the BH Hawking temperature will become higher than the CBR temperature. The BH will be in a stationary state given by Eq. (113). Then, the BH will start to emit Hawking radiation and one assumes a first emission from the BH excited state corresponding to the BH principal quantum number $n = m$, which has a total energy $E_m = -\frac{1}{2}\sqrt{m}$, given by Eq. (69) and a corresponding total mass $M_m = 2\sqrt{m}$, given by Eq. (68), to a state with, say $n = m_1 < m$, which corresponds to a total energy $E_{m_1} = -\frac{1}{2}\sqrt{m_1}$ and to a corresponding total mass $M_{m_1} = 2\sqrt{m_1}$. It will be $M_{m_1} < M_m$ because the infalling particle in Hawking's mechanism of particle creation by BH has negative mass. The energy jump between the two levels is

$$\Delta E_{m\to m_1} \equiv E_{m_1} - E_m = \frac{1}{2}\left(\sqrt{m} - \sqrt{m_1}.\right) \tag{123}$$

In other words, the energy of the first absorbed particle having negative mass is transferred, together with its part of the wave function, to the two-particle system governed by the equations from Eq. (63) to Eq. (66), which is now entangled with the emitted particle. By using Eq. (122) and setting $n = m_1$ one finds that the part of the wave function in the interior of the horizon, i.e., the part of the wave function associated with the particle having negative mass (infalling mode), which has been transferred to the two particle system, is

$$|\psi_{(m\to m_1)}(r,t) >\equiv \exp\left(-iE_{m_1}t\right)|\varphi_{m_1}(r) > - \exp\left(-iE_m t\right)|\varphi_m(r) > . \tag{124}$$

Now, let us consider a second emission, which corresponds to the transition from the state with $n = m_1$ to a state with, say, $n = m_2 < m_1$. The BH total energy changes from $E_{m_1} = -\frac{1}{2}\sqrt{m_1}$ to $E_{m_2} = -\frac{1}{2}\sqrt{m_2}$ while the BH mass changes from $M_{m_1} = 2\sqrt{m_1}$ to $M_{m_2} = 2\sqrt{m_2} < M_{m_1}$. The energy jump between the two levels is

$$\Delta E_{m_1\to m_2} \equiv E_{m_2} - E_{m_1} = \frac{1}{2}\left(\sqrt{m_1} - \sqrt{m_2}\right). \tag{125}$$

The energy of the second absorbed particle having negative mass is transferred, together with its part of the wave function, again to the two-particle system governed by the equations from Eq. (63) to Eq. (66), which is now entangled with both the emitted particles. By using again Eq. (122) and setting $n = m_2$ and $m = m_1$, one finds that the part of the wave function of the second infalling mode, which has been transferred to the two particle system is

$$|\psi_{(m_1 \to m_2)}(r,t) > \equiv \exp\left(-iE_{m_2}t\right)|\varphi_{m_2}(r) > - \exp\left(-iE_{m_1}t\right)|\varphi_{m_1}(r) > . \tag{126}$$

Let us consider a third emission, which corresponds to the transition from the state with $n = m_2$ to a state with, say, $n = m_3 < m_2$. The BH total energy changes from $E_{m_2} = -\frac{1}{2}\sqrt{m_2}$ to $E_{m_3} = -\frac{1}{2}\sqrt{m_3}$ while the BH mass changes from $M_{m_2} = 2\sqrt{m_2}$ to $M_{m_3} = 2\sqrt{m_3} < M_{m_2}$. The energy jump between the two levels is

$$\Delta E_{m_2 \to m_3} \equiv E_{m_3} - E_{m_2} = \frac{1}{2}\left(\sqrt{m_2} - \sqrt{m_3}.\right) \tag{127}$$

The energy of the third absorbed particle having negative mass is transferred, together with its part of the wave function, again to the two-particle system governed by the equations from Eq. (63) to Eq. (66), which is now entangled with the 3 emitted particles. By using again Eq. (122) and setting $n = m_3$ and $m = m_2$, one finds that the part of the wave function of the second infalling mode, which has been transferred to the two particle system is

$$|\psi_{(m_2 \to m_3)}(r,t) > \equiv \exp\left(-iE_{m_3}t\right)|\varphi_{m_3}(r) > - \exp\left(-iE_{m_2}t\right)|\varphi_{m_2}(r) > . \tag{128}$$

The process will continue again, and again, and again... till the *Planck distance* and the *Planck mass* are approached by the evaporating BH. At that point, as it has been stressed in previous Sections, the GUP prevents the total BH evaporation in exactly the same way that the HUP prevents the hydrogen atom from total collapse [72] and the two-particle system governed by the equations from Eq. (63) to Eq. (66) is entangled with all the Hawking quanta emitted at that time. Hence, the BH arrives at a minimum energy level which is compatible with the GUP. Such an energy level corresponds to Eq. (98) and consists of a negative energy having the absolute value of the Planck energy. In other words, subsequent "absorptions" by the BH of the wave functions of absorbed particles having negative mass-energy in Hawking's mechanism of particle creation lead the BH total wave function which at the starting of the emission process was given by Eq. (113), to evolve till the wave-function of the BH minimum energy level

$$|\psi_1(r,t) > = \exp\left(-iE_1 t\right)|\varphi_1(r) > . \tag{129}$$

Clearly, the evolution of BH evaporation that it has been discussed above is *unitary*.

Therefore, it has been shown that SBHs, which are considered the fundamental bricks of quantum gravity, are well defined quantum mechanical systems, having ordered, discrete quantum spectra, which preserve physical information by restoring predictability in gravitational collapse.

Conclusion

A novel approach to black hole (BH) quantum gravity, which has been recently developed by the Author and Collaborators in the series of papers [27-42] and [45–48], has been carefully reviewed and refined. Such an approach shows that the SBH is a quantum non-singular two-particle system where the two components, the "nucleus" and the "electron", strongly interact with each other through a quantum gravitational interaction. After summarizing the general quantum gravity problem, the analysis started from the semi-classical Bohr-like approach to BH quantum physics which was early developed by the Author and Collaborators in [27-42]. After that, by applying Feynman's path integrals fapproach [75] to the historical Oppenheimer and Snyder gravitational collapse [50] and by setting the constraints for the formation of the SBH, the gravitational potential, the Schrodinger equation and the solution for the energy levels for the SBH has been derived. Remarkably, the obtained results are consistent with previous ones in the literature. Therefore, the obtained results have shown that the SBH is the gravitational analogous of the hydrogen atom and this could, in principle, drive to a space-time quantization based on a quantum mechanical particle approach. Following this analogy with the hydrogen atom, the de Broglie hypothesis has been used in the SBH context on the wave nature of the BH "electron". This means that such a particle does not orbit the nucleus in the same way as a planet orbits the Sun, but instead exists as a standing wave. The correct analogy is that of a large and often oddly shaped "atmosphere" (the BH "electron"), distributed around a relatively tiny planet (the BH "nucleus"). The correct physical interpretation of such an "atmosphere" is nothing else than the BH horizon modes. By following with caution the analogy between the SBH Schrodinger equation and the traditional Schrodinger equation of the s states ($l = 0$) of the hydrogen atom, the SBH Schrodinger equation has been solved and discussed in detail. The approach permitted us to find the quantum gravitational quantities which are the gravitational analogous of the fine structure constant and of the Rydberg constant. An intriguing result has been that such quantities are not constants. Instead, they are dynamical quantities having well defined discrete spectra. In particular, the spectrum of the "gravitational fine structure constant" is exactly the set of non-zero natural numbers $\mathbb{N} - \{0\}$. Therefore, the important consequence is that the SBH results to be a well defined quantum gravitational system, which obeys Schrodinger's theory: the "gravitational hydrogen atom". The area quantum and the quantum representation of the SBH ground state at the Planck scale have been also derived. After that it has been also shown that the SBH quantization permits to compute the SBH entropy as a function of the BH principal quantum number in terms of Bekenstein-Hawking entropy and 3 sub-leading corrections. The coefficient of the formula of Bekenstein-Hawking has been checked through a thermodynamic analysis by finding that it is reduced to a quarter of its standard value. After that, Hawking radiation has been analysed by discussing its connection with the BH quantum structure. Finally, it has been shown that, remarkably, the time evolution of the "gravitational hydrogen atom", which is governed by a time-dependent Schrodinger equation, solves the BH information paradox by restoring predictability in gravitational collapse.

Acknowledgments

The Author thanks Prof. Arun Kenath for inviting him to write this Book Chapter on BH physics.

References

[1] A. Einstein, *Preuss. Akad. Wiss.* Berlin, Sitzber., **778** (1915).

[2] L. Landau and E. Lifsits, *Classical Theory of Fields* (3rd ed.), London: Pergamon (1971).

[3] C. Corda, *Int. Journ. Mod. Phys. D*, **18**, 2275 (2009).

[4] A. Pais, *Subtle Is the Lord: The Science and the Life of Albert Einstein*, Oxford University Press (2005).

[5] C. Corda, *New Adv. Phys.* **7,** 67 (2013).

[6] L. H. Ryder, *Quantum Field Theory*, 2nd Edition, Cambridge University Press (1996).

[7] C. W. Misner, K. S. Thorne and J. A. Wheeler, *Gravitation* (W. H. Feeman and Co., 1973).

[8] K. O. Friedrichs, *Mathematical aspects of the Quantum Theory of Fields*. Interscience Publishers (1953).

[9] J. Bekenstein, *Lett. Nuovo Cimento*, **11**, 467 (1974).

[10] S. W. Hawking, Commun. Math. Phys. **43**, 199 (1975).

[11] J. D. Bekenstein, in *Proceedings of th Eight Marcel Grossmann Meeting*, T. Piran and R. Ruffini, eds., pp. 92-111 (World Scientific Singapore 1999).

[12] S. Hod, *Phys. Rev. Lett.* **81**, 4293 (1998).

[13] V. Mukhanov, *JETP Letters* **44**, 63 (1986).

[14] J. D. Bekenstein and V. Mukhanov, *Phys. Lett. B* **360**, 7 (1995).

[15] S. Das, P. Ramadevi and U. A. Yajnik, *Mod. Phys. Lett. A* **17**, 993 (2002).

[16] A. Barvinski, S. Das, G. Kunstatter, *Phys. Lett. B* **517**, 415 (2001).

[17] L. Modesto, J. W. Moffat, P. Nicolini, *Phys. Lett. B* **695**, 397 (2011).

[18] C. Kiefer and T. Schmitz, *Phys. Rev. D* **99**, 126010 (2019).

[19] P. Hajicek and C. Kiefer, *Nucl. Phys. B* **603**, 531 (2001).

[20] P. Hajicek and C. Kiefer, *Int. J. Mod. Phys. D* **10**, 775 (2001).

[21] A. Saini and D. Stojkovic, *Phys. Rev. D* **89**, 044003 (2014).

[22] E. Greenwood and D. Stojkovic, *JHEP* **0806**, 042 (2008).

[23] J. E. Wang, E. Greenwood and D. Stojkovic, *Phys. Rev. D* **80**, 124027 (2009).

[24] A. Davidson, *Phys. Rev. D* **100**, 081502 (2019).

[25] A. Davidson, *Phys. Lett. B* **780**, 29 (2018).

[26] M. Maggiore, *Nucl. Phys. B* **429**, 205 (1994).

[27] C. Corda, *J. High Energ. Phys.* **2011**,101 (2011).

[28] C. Corda, *Int. Jour. Mod. Phys. D* **21**, 1242023 (2012).

[29] C. Corda, *Eur. Phys. J. C* **73**, 2665 (2013).

[30] C. Corda, *EJTP* **11**, 27 (2014).

[31] C. Corda, *Ann. Phys.* **353**, 71 (2015).

[32] C. Corda, *Class. Quantum Grav.* **32,** 195007 (2015).

[33] C. Corda, *Adv. High En. Phys.* **2015**, 867601 (2015).

[34] C. Corda, *AIP Conf. Proc.* **1648**, 020004 (2015).

[35] C. Corda, *Int. Journ. Theor. Phys.* **54**, 3841 (2015).

[36] C. Corda, *Mod. Phys. Lett. A*, 33, 1850069 (2018).

[37] C. Corda, S. H. Hendi, R. Katebi, N. O. Schmidt, *J. High Energ. Phys.* **2013**, 8 (2013).

[38] C. Corda, S. H. Hendi, R. Katebi, N. O. Schmidt, *Adv. High En. Phys.* **2014**, 527874 (2014).

[39] C. Corda, S. H. Hendi, R. Katebi, N. O. Schmidt, *Adv. High En. Phys.* **2014**, 530547 (2014).

[40] S. Haldar, C. Corda and S. Chakraborty, *Adv. High En. Phys.* **2018**, 9851598 (2018).

[41] Y. Heydarzade, H. Hadi, C. Corda, F. Darabi, *Phys. Lett. B* **776**, 457 (2018).

[42] L. Crowell and C. Corda, *Entropy* **22(3)**, 301 (2020).

[43] N. Bohr, *Philos. Mag.* **26**, 1 (1913).

[44] N. Bohr, *Philos. Mag.* **26**, 476 (1913).

[45] C. Corda and F. Feleppa, *Adv. Theor. Math. Phys.* (2022), pre-print in arXiv:1912.06478.

[46] C. Corda, F. Feleppa and F. Tamburini, *EPL* **132**, 30001 (2020).

[47] C. Corda, arXiv:2012.08967 (2020).

[48] C. Corda, F. Feleppa, F. Tamburini and I. Licata, arXiv:2104.05451 (2021).

[49] N. Rosen, *Int. Journ. Theor. Phys.* **32**, 8, (1993).

[50] J. R. Oppenheimer and H. Snyder, *Phys. Rev.* **56**, 455 (1939).

[51] L. de Broglie, *Ann. de Physique* **(10)3**, 22 (1925).

[52] W. H. Press, *Astrophys. J.* **170**, L105 (1971).

[53] M. Maggiore, *Phys. Rev. Lett.* **100**, 141301 (2008).

[54] E. Spallucci, A. Smailagic, *Phys. Lett B* **816**, 136180 (2021).

[55] M. K. Parikh, *Gen. Rel. Grav.* **36**, 2419 (2004).

[56] M. K. Parikh and F. Wilczek, *Phys. Rev. Lett.* **85**, 5042 (2000).

[57] S. W. Hawking, *Phys. Rev. D* **14**, 2460 (1976).

[58] C. Corda, *Ann. Phys.* **337**, 49 (2013), definitive version with corrected typos in arXiv:1305.4529v3.

[59] A. E. Roy and D. Clarke, *Astronomy: Principles and Practice*, Fourth Edition (2003, PBK).

[60] S. W. Hawking, “The Path Integral Approach to Quantum Gravity”, in *General Relativity: An Einstein Centenary Survey*, eds. S. W. Hawking and W. Israel, (Cambridge University Press, 1979).

[61] R. Banerjee and B. R. Majhi, *Phys. Lett. B* **674**, 218 (2009).

[62] L. Motl, *Adv. Theor. Math. Phys.* **6**, 1135 (2003).

[63] J. York Jr., *Phys. Rev. D* **28**, 2929 (1983).

[64] N. Bohr, *Zeits. Phys.* **2**, 423 (1920).

[65] B. Zhang, Q.-Y. Cai, L. You,, and M. S. Zhan, *Phys. Lett. B* **675**, 98 (2009).

[66] B. Zhang, Q.-Y. Cai, M. S. Zhan, and L. You, *Ann. Phys.* **326**, 350 (2011).

[67] B. Zhang, Q.-Y. Cai, M. S. Zhan, and L. You, *EPL* **94**, 20002 (2011).

[68] B. Zhang, Q.-Y. Cai, M. S. Zhan, and L. You, *Int. Journ. Mod. Phys. D* **22**, 1341014 (2013).

[69] H. P. Nollert, *Phys. Rev. D* **47**, 5253 (1993).

[70] N. Andersson, *Class. Quant. Grav.* **10**, L61 (1993).

[71] L. Motl and A. Neitzke, *Adv. Theor. Math. Phys.* **7**, 307 (2003).

[72] R. J. Adler, P. Chen and D. I. Santiago, *Gen. Rel. Grav.* **33**, 2101 (2001).

[73] D. L. Beckerdoff and C. W. Misner, D. L. Beckerdoff's A. B. Senior Thesis, Princeton Univeristy (1962).

[74] C. Corda and H. J. Mosquera Cuesta, *Mod. Phys. Lett. A* **25**, 2423 (2010).

[75] R. Feynman, *Feynman's Thesis – A New Approach to Quantum Theory*, Edited by Laurie M Brown, World Scientific (2005).

[76] A. Messiah, *Quantum Mechanics, Vol. 1*, North-Holland, Amsterdam (1961).

[77] B. G. Schmidt, *Gen. Rel. Grav.* **1**, 269 (1971).

[78] A. Kempf, G. Mangano, R. B. Mann, *Phys. Rev. D* **52**, 1108 (1995).

[79] H. Kleinert, *Path Integrals in Quantum Mechanics, Statistics, Polymer Physics, and Financial Markets*, 5th edition, World Scientific, Singapore (2009).

[80] N. Zettili, *Quantum Mechanics: Concepts and Applications* (2nd ed.). Chichester: Wiley (2009).

[81] F. Scardigli, R. Casadio, *Eur. Phys. J. C* **75**, 425 (2015).

[82] A. Farag Ali, *Class. Quantum Grav.* **28**, 065013 (2011).

[83] A. Galiautdinov, Lewis H. Ryder, *Gen. Relativ. Gravit.* **49**, 82 (2017).

[84] V. M. Tkachuk, *Found. Phys.* **46 (12)**, 1666 (2016).

[85] H. Abele and H. Leeb, *New J. Phys.* **14**, 055010 (2012).

[86] Z. W. Feng, S. Z. Yangb, H. L. Li and X. T. Zu, *Phys. Lett. B* **768**, 81 (2017)

[87] G. Lambiase, F. Scardigli, *Phys. Rev. D* **97**, 075003 (2018).

[88] F. Scardigli, *J. Phys. Conf. Ser.* **1275**, 012004 (2019).

[89] J. D. Bekenstein, *Phys. Rev. D* **7**, 2333 (1973).

[90] Hao-Peng Yan, Wen-Biao Liu, *Phys. Lett. B* **759**, 293 (2016).

[91] G. Scharf, *Nuovo Cim. B* **113**, 821 (1998).

[92] K. V. Krasnov, *Class. Quant. Grav.* **16**, 563 (1999).

[93] Y. Yoon, *Jour. Kor. Phys. Soc.* **68**, 730 (2016).

[94] S. W. Hawking, *Phys. Rev. D* **72**, 084013 (2005).

[95] L. Susskind, *The Black Hole War: My Battle with Stephen Hawking to Make the World Safe for Quantum Mechanics*, Little, Brown and Company (2008).

[96] S. W. Hawking, arXiv:1401.5761 (2014).

[97] J. J. Sakurai, *Modern Quantum Mechanics*, Pearson Education (2006).

[98] G. 't Hooft, *Int. J. Mod. Phys. A* **11**, 4623 (1996).

[99] Samir D. Mathur, *Pramana* **79**, 1059 (2012).

Chapter 7

Black Hole Menagerie, Charged/Dyonic BHs and Radiation from Interacting Dyonic BH Pairs

Patrick Das Gupta* **and Mohd. Sirtaz**†
Department of Physics and Astrophysics,
University of Delhi, Delhi, India

Abstract

Although black holes (BHs) as well as gravitational waves (GWs) were predicted from general relativity more than hundred years back, it is only in recent times that substantial observational evidence is mounting up concerning their actual existence, thanks to LIGO/VIRGO detectors, Event Horizon telescope and tracking of stellar motion near the Galactic Centre. In this article, after discussing basic physics concerning classical BHs, we provide a pedagogical exposition of supermassive black holes, their relevance to the understanding of active galactic nuclei, formation of primordial black holes as well as quantum aspects and challenges associated with BHs.

Thereafter, we describe charged BHs, Penrose process for energy extraction from Kerr BHs and Wald's proposal concerning a Kerr BH slowly becoming a Kerr-Newman BH in the presence of a uniform magnetic field. In the context of BHs bearing magnetic charge, we discuss both

*Corresponding Author's Email: pdasgupta@physics.du.ac.in
†Corresponding Author's Email: sirtaz@physics.du.ac.in

In: A Guide to Black Holes
Editor: Kenath Arun
ISBN: 979-8-88697-163-7

magnetic monopoles as well as dyons, and their emergence from various models like string theory, GUTs and electroweak theories, etc.

In the later portions, we concentrate on our recent research work pertaining to the non-relativistic dynamics of dyon-dyon interaction that includes mutual gravitational attraction. From the derived classical equations of motion, we obtain not only the well known Schwinger-Zwanziger quantization condition for dyons using Saha's argument based on quantized angular momentum of electromagnetic field but also a scalar virial theorem for an astrophysical system consisting of point particles, some of which carry both electric and magnetic charges.

In the final sections, we obtain expressions for the generated electromagnetic wave as well as gravitational wave amplitudes, and the corresponding luminosities due to dyon-dyon interactions. Lastly, we discuss the results after computing these quantities using a range of values for the mass, electric and magnetic charges, etc. of the dyonic BHs.

1. Introduction

It is well established that special relativity not only placed time and space almost on the same footing, it also brought energy and momentum together in a single framework, entailing mass and energy to be identical, modulo a factor of c^2. Einstein's general relativity (GR), a relativistic theory of gravitational interactions, that rests firmly on the pillars of equivalence principle and special relativity, had gone one step ahead and demonstrated that gravity is simply a manifestation of space-time geometry.

Einstein's equivalence principle states that in a sufficiently 'small' freely falling frame of reference, gravity disappears for a sufficiently 'small' duration in this inertial frame, irrespective of the detailed nature of the gravitating source that is causing the free fall. The 'smallness' of such a local inertial frame, of course, depends on how strong, time varying and non-uniform the local gravity (i.e., local space-time geometrical curvature) is. The space-time geometry in GR is determined by the energy and momentum densities as well as their flux densities corresponding to all matter, radiation and any other form of gravitating sources present in the universe.

One of the signature predictions of GR is the existence of black hole (BH) solutions. A classical BH solution is characterized by an event horzon (EH) that surrounds the space-time singularity and acts as a one-way boundary for all null-like and time-like geodesics. The EH is analogous to a trapdoor that leads to an unending abyss. By characterizing space-time singularity to be a point where all geodesics get terminated (i.e., become incomplete), Penrose had proved a powerful singularity theorem [1] that had made use of the focussing effect of geodesics that ensue from the Raychaudhari equation due to matter that satisfy strong energy condition [2]. This theorem paved the way for the acceptance of singularities resulting from any kind of gravitational collapse involving standard matter.

A BH singularity is hidden by an EH, a mathematical surface encompassing the singularity. Any object that crosses the EH towards the singular point cannot retrace its footsteps back or come out. For an observer outside of the EH, time stands still on this one-way boundary and any radiation emitted outward from this surface suffers infinite redshift, as long as one ignores quantum physics (for an extremely lucid introduction to BHs, see [3]). The idea that BHs are not merely mathematical solutions of GR but are generically created as end products of stellar evolution in the universe, emerged essentially from the pioneering research studies of Subrahmanyan Chandrasekhar (for pedagogical expositions, see e.g., [4], [5], [6]). From theoretical studies, the present understanding is that stellar mass BHs result from the collapse of stellar cores heavier than $\sim 3 - 10\ M_\odot$.

The simplest BH case ensues from an exact solution of the Einstein equation, corresponding to the exterior of a spherically symmetric mass distribution of finite radius r_0, obtained by Karl Schwarzschild in 1915, that was published in 1916, not long before he passed away ([7]). For $r > r_0$, as shown by Schwarzschild, the line-element for this exact solution has the following form,

$$ds^2 = [1 - f(r)]c^2dt^2 - \frac{dr^2}{1 - f(r)} - d\theta^2 - \sin^2\theta\, d\phi^2 \tag{1}$$

with $f(r)$ given by the expression,

$$f(r) = \frac{R_s}{r} \ , \tag{2}$$

where the Schwarzschild R_s radius is defined as,

$$R_s \equiv \frac{2GM}{c^2} \cong 3 \text{ km} \left(\frac{M}{M_\odot}\right) , \tag{3}$$

M being the total mass of the spherically symmetric matter distribution.

The Schwarzschild BH solution, with all its strangeness, emerges from the above expressions naturally, when the spherically symmetric object of mass M shrinks to a radius $r_0 < R_s$. In such a situation, as we shall argue in the next section, the entire matter making up the object collapses unhaultingly from r_0 to $r = 0$. Hence, matter density as well as the space-time curvature become infinite at $r = 0$, making the point a singularity. The EH corresponds to the surface $r = R_s$ with its size being characterized by the Schwarzschild radius R_s. As can be seen readily, from eqs.(1) and (2), the metric components g_{tt} and g_{rr} become 0 and $-\infty$, respectively, on the EH.

In the realm of classical physics, a BH is uniquely characterized by its mass, angular momentum and charge (see e.g., [8], and the references therein). It does not matter whether a BH is created from the gravitational collapse of neutrons, anti-neutrons, photons, neutrinos/anti-neutrinos or any other electrically neutral particle, as it will remember in this case, only the total mass and the total angular momentum that went in to form the BH, ignoring blissfully the baryon number or the lepton number of the initial parent object. This follows from the standard 'no hair' theorem for classical BHs.

In other words, unlike the macroscopic objects we encounter in our day today lives, classical BHs are completely specified by just three physical quantities: M, $\vec{L}$ and Q. Furthermore, as far as GR is concerned, BHs do not have any non-spacetime or material structure whatsoever. Mass, angular momentum and charge appear only as parameters in the space-time metric of the BHs, that determines the EH and the singularity. This is precisely the reason why Chandrasekhar had stated in his tome on BHs [8]:

'The black holes of nature are the most perfect macroscopic objects there are in the universe: the only elements in their construction are our concepts of space and time. And since the general theory of Relativity provides only a single unique family of solutions for their descriptions, they are the simplest objects as well.'

However, by considering quantum states of gravitons corresponding to BH exterior regions, it has been argued cogently in a very recent paper that BHs do possess 'quantum hairs', notwithstanding the above classical view of BHs [9].

In the next section, we begin with basic physics related to BHs and highlight their role in explaining the origin of very high luminosities from active galactic nuclei like quasars and blazars. Supermassive black holes as well as sub-solar mass primordial BHs, along with the puzzles and challenges associated with them, have been discussed. We also dwell briefly on the subject of 'Black hole atoms', Hawking radiation and the information loss paradox, in this section.

Section III describes charged BHs and Wald's proposed mechanism on how a Kerr BH, embedded in a uniform magnetic field, can get electrically charged. Thereafter, the topic of magnetic monopoles and dyons has been discussed. Plausible origins of these hypothetical particles in string theory, Yang-Mills theories, Grand Unified theories (GUTs) as well as electroweak theory have been highlighted.

Our recent research work on the non-relativistic dynamics of dyon-dyon interaction has been taken up in section IV. Apart from the electromagnetic forces between the dyons, Newtonian gravity between them has also been incorporated. In this section, we provide a detailed mathematical derivation of the relevant dynamical equations as well as the ensuing constants of motion, for a pair of dyons. Saha's method of charge quantization in the presence of a magnetic monopole has been extended to the case of two static dyons in order to obtain the Schwinger-Zwanziger quantization condition. Furthermore, a scalar virial theorem for a large gravitating system of point particles that includes a collection of dyons has been proved, in this section.

Section V covers the physics of field propagation in relativistic theories, in general, as well as the topic of gravitational radiation in general relativity. Electromagnetic radiation and gravitational waves resulting from dyon-dyon interactions have been elaborately dealt with in this section.

Finally, in section VI, quantitative estimates of the electromagnetic wave and gravitational wave amplitudes, as well as the corresponding luminosities have been computed by considering various values for the mass, electric and magnetic charges of an interacting electrically charged BH and a magnetically charged BH. Interestingly enough, for a given set of parameters, the EM radiation from such a pair of BHs mimics that of a Fast Radio Burst.

2. Black Hole Diversity: Supermassive, Primordial and Quantum Effects

2.1. BHs: Physical Aspects

Resuming our discussion from the previous section on the Schwarzschild solution, associated with a spherically symmetric mass, eqs.(1) and (2) imply that if the radius r_0 of this sphere is less than R_s then a test particle in its vicinity cannot be at a constant radius r with $r_0 < r < R_s$. This is because, for a constant r (i.e., $dr = 0$) beneath the EH,

$$ds^2 = [1 - \frac{R_s}{r}]c^2dt^2 - d\theta^2 - \sin^2\theta\, d\phi^2 < 0 \quad , \tag{4}$$

entailing any 'constant r' world-line with $r < R_s$ to be space-like. This prohibits any physical particle to be at a constant r below R_s.

In other words, even the surface matter of this spherical object (and thereby, the entire object) cannot be at a constant r_0, and instead, must contract incessantly. This gravitating ball cannot expand as that would violate energy conservation (see the discussion in the following paragraphs). Therefore, from the Schwarzschild solution, it is evident that a spherically symmetric object of initial radius less than R_s necessarily shrinks inexorably to the singular point $r = 0$.

It is a standard result in GR that if a space-time geometry admits a Killing vector ξ^μ then the quantity $\xi_\mu p^\mu$ is a constant of motion for a test particle having a 4-momentum p^μ. For the Schwarzschild space-time described by eqs. (1) and (2), it is obvious that $\xi^\mu = \delta^\mu_0$ is a time-like Killing vector since the metric is invariant under $t \to t$+constant. Hence, for a test particle of rest mass m, initially at rest, at the radial coordinate r,

$$E = \xi_\mu p^\mu = g_{\mu\nu}\xi^\nu p^\mu = g_{00}p^0 = \sqrt{1 - R_s/r}\, mc^2 \tag{5}$$

is its conserved energy. From the above expression, at distances $r \gg R_s$, one finds that the conserved energy, $E \approx (1 - R_s/2r)\, mc^2 = mc^2 - GMm/r$, takes the expected form, of being the sum of the test particle's rest energy and its gravitational potential energy.

According to eq. (5), farther away the particle is located, greater is its energy. Thus, a particle at rest initially at r_i can only fall inwards with $\dot{r}$ increasing with time, and not move upwards in the absence of an external force acting on it. Hence, a pressureless ball of dust-like matter can only contract. It is also interesting to note from the above equation that one may extract almost the entire rest energy mc^2 of a test particle by lowering it ever so slowly from $r_i \gg R_s$ down to the event horizon of a Schwarzschild BH since at $r \approx R_s$, $E \approx 0$.

Therefore, very large luminosities and rapid time variability exhibited by the high frequency EM radiation emitted by the active galactic nuclei (AGNs) (e.g., blazars and quasars) can very effectively be explained by invoking infall of matter from far away to regions close to the EH surface of a supermassive black hole (SMBH). In fact, causality argument itself implies that fluctuations in the accretion processes taking place close to the event horizon of the SMBH can give rise to variability time scales $\Delta t \sim \kappa\ R_s/c$, where κ is a dimensionless parameter $\gtrsim 1$. For instance, typically observed $\Delta t \sim$1 hour variability time scales in a blazar would mean $R_s \sim 3.6 \times 10^{14}/\kappa$ cm, implying a SMBH of mass $\sim 10^8 M_\odot$.

Similarly, one may estimate from eq. (5) the maximum luminosity possible from matter forming an accretion disc, and trickling down from an outer radius $\gg R_s$ slowly onto a Schwarzschild BH due to viscous dissipation at a rate $\dot{m}$,

$$L_{max} \cong \dot{m}c^2 \left[1 - \sqrt{1 - \frac{R_s}{r_0}}\,\right] \approx 0.18\,\dot{m}c^2 \tag{6}$$

where radius $r_0 = 3R_s$ is the innermost stable circular orbit in the disc (for very brief introductions to BH accretion, see e.g., [6], [10]). From eq. (6), one draws the conclusion that BH accretion can result in higher efficiency of conversion of matter into radiation than nuclear fission or fusion.

2.2. Supermassive Black Holes

So far, most galaxies that have been studied closely reveal presence of SMBHs with mass $\sim 10^6 - 10^7\ M_\odot$ at their centres. Very heavy SMBHs, on the other hand, are associated with power house of AGNs. This is because, from eq. (6), high luminosity requires large values of accretion rate $\dot{m}$. But higher the luminosity, greater is the radiation pressure on the infalling matter prohibiting the latter to be accreted at a rate more than the Eddington rate. Hence, in general, the AGN power is limited by the Eddington luminosity which is proportional to the mass of the feeding BH, since larger the BH mass greater is its inward gravitational force on the matter to counteract the outward force due to radiation pressure. Therefore, it is natural to expect a correlation between AGN luminosity and the mass of the underlying SMBH.

The heaviest SMBH detected, as of now, lies in the quasar TON 618, weighing $\sim\ 66 \times 10^9\ M_\odot$. The light from this quasar that we see was emitted $\sim 10.4 \times 10^9$ yrs ago, when the universe was only $\sim 3 \times 10^9$ yrs old. So, how are such SMBHs created in the universe so early in latter's history? This is one of the biggest conundrums of cosmology. Several models have been proposed to address this puzzle (for a review, see e.g., [11]).

Interestingly enough, if dark matter (DM) candidates instead of being somewhat massive particles with rest energy more than 100 GeV, are actually ultralight axion-like pseudoscalars with rest energy $m_{DM}c^2 \lesssim 10^{-20}$ eV, then a fraction of galactic mass scale DM halo can undergo Bose-Einstein (BE) condensation. The quantum evolution of such BE condensates, with self-gravity included, can naturally lead to creation of ultramassive BHs of mass M_{SMBH}, satisfying a condition,

$$m_{DM}\, M_{SMBH} \gtrsim m_{Pl}^2 \Rightarrow \left(\frac{m_{DM}c^2}{10^{-20}\ \mathrm{eV}}\right)\left(\frac{M_{SMBH}}{10^{10}\, M_\odot}\right) \gtrsim 1\ , \tag{7}$$

when the universe is barely billion years old [12, 13, 14]. In eq. (7), $m_{Pl} \equiv \sqrt{\hbar c/G} \approx 10^{-5}$ gm is the Planck mass.

Origin of ultraheavy SMBHs when the universe is relatively young is enigmatic, as far as our theoretical understanding is concerned. On the other hand, creation of sub-solar mass BHs (or primordial BHs) in the radiation-dominated era of our universe had been theoretically predicted about five decades ago (see [15, 16], and also [17] for a related idea).

2.3. Primordial Black Holes

The argument delineated by Hawking and Carr as to why primordial BHs (PBHs) are expected to be formed in the early universe is remarkably simple [15, 16]. At any epoch, the causally connected regions in an expanding universe is limited by the Hubble radius, $R_H(t) \sim c/H(t)$, where the Hubble parameter is given by $H(t) = \dot{a}/a$, $a(t)$ being the scale factor corresponding to a homogenous and isotropic cosmic model and $\dot{a} \equiv da/dt$ (for an elementary exposition, see e.g., [18]). In the radiation-dominated era, the scale factor goes as $a(t) \propto t^{1/2}$ so that $H(t) \sim 1/t$ and $R_H(t) \sim c\, t$, while the highly relativistic particles that are in thermodynamic equilibrium have a temperature $T(t) \propto 1/a(t) \propto t^{-1/2}$.

Therefore, the energy density of the dominant components of the cosmic content is given by,

$$\rho(t)c^2 \sim a_B T^4 \sim \frac{3c^2}{8\pi G t^2} \tag{8}$$

The mass enclosed within a Hubble radius then follows from eq. (8),

$$M_H(t) \sim \frac{4\pi R_H^3}{3}\rho(t) \sim \frac{4\pi}{3}(ct)^3\rho(t) \sim \frac{c^3 t}{2G} = 2\times 10^3\left(\frac{t}{10^{-35}\ s}\right)\ \mathrm{gm} \tag{9}$$

From eq. (9), the Schwarzschild radius associated with $M_H(t)$ is,

$$R_s^H(t) \equiv \frac{2GM_H(t)}{c^2} \sim \frac{2G}{c^2}\frac{c^3 t}{2G} = ct \sim R_H(t)\ ! \tag{10}$$

The scale R_H, characterizing the size of a causally connected region in an expanding universe, while R_s providing the minimum size of a gravitating object below which it necessarily collapses into a BH, are two independent scales. So, it is rather surprising that they turn out to be comparable, as seen from eq. (10). In other words, if at time t, due to statistical fluctuations in the thermal energy density of the relativistic components within a Hubble radius, the mass becomes slightly greater than $M_H(t)$ of eq. (9), the entire contents within the Hubble radius would collapse to form a PBH.

Of course, there are many other theoretical models proposed in the literature concerning creation of PBHs in the early universe e.g., collision of bubble walls due to first order phase transitions, collapse of trapped false vacuum pockets, gravitational collapse of inflaton/scalar fields/dark matter/dark energy, etc. (see e.g., [19, 20, 21, 22, 23], and the references therein.).

Based on the observed extragalactic gamma-ray background as well as constraints arising from the Big Bang nucleosynthesis and the observed spectrum of the cosmic microwave background, stringent restrictions on the mass fraction as a function of the PBH mass ensue (for a recent review, see [19]). Here is an irony: Observed super heavy SMBHs are fairly ubiquitous, although their origin so early in the cosmic history is still shrouded in mystery. PBHs, on the other hand, have so far evaded detection inspite of the well understood scenarios that would entail their formation.

2.4. Black Holes and Quantum Theory

Due to the predicted Hawking temperature of BHs [24],

$$T_H(M) = \frac{\hbar c}{4\pi k_B R_s} = \frac{m_{Pl}^2 c^2}{8\pi k_B M} \sim \frac{10^{26}}{M \text{ (in gm)}} \; {}^\circ K \, , \tag{11}$$

associated with the semi-classical Hawking radiation resulting from the quantum effects of matter fields near the EH of a classical BH of mass M, PBHs of mass $10^{14} - 10^{15}$ gm created more than billion years ago would be evaporating today, emitting gamma-ray photons copiously. Moreover, PBHs created before the electroweak phase transition may also resolve the issue of baryogenesis in the early universe provided there are adequate CP-violations in the decay of GUT bosons (see e.g., [21, 22], and the references therein.).

Now, through a gravitational capture in the early universe, a PBH of mass M can certainly form a quantum mechanical bound state with an electron or a proton of mass m, with hydrogen-atom-like energy levels,

$$E_n \approx -\frac{G^2 M^2 m^3}{2\hbar^2 n^2} = -\frac{mc^2}{2n^2}\left(\frac{mc^2}{8\pi k_B T_H(M)}\right)^2, \tag{12}$$

that follow from the energy eigenstates analysis and eq. (11). From eq. (12), it is easy to show that such a 'Black hole atom' is unstable due to the ionizing effects of its own Hawking radiation [10].

Non-stop Hawking evaporation of a BH until the latter's mass approaches Planck mass m_{Pl}, when quantum gravity effects are expected to take over, leads to the well known BH information loss paradox [24, 25, 26]. This is essentially due to the fact that a pure quantum state $|\psi_0^{BH}>$ describing a BH to begin with, becomes a mixed state that represents the thermal quantum state of the emitted particles due to Hawking evaporation. Quantum theory's unitary evolution of state-vectors strictly forbids a pure state evolving into a mixed state.

An elegant resolution of this paradox has been developed by Samir Mathur and his group based on the string theory paradigm by demonstrating that for no microstate describing a BH in string theory, a conventional EH exists and that, the usual BH space-time ends just before the EH. Rather the microstates describing a stringy BH give rise to a 'fuzzy' horizon, and the Hawking emission from such a 'fuzzball' corresponds to a pure state, without violating unitarity, not unlike the radiation from a thermally hot entity [27, 28, 29, 30, 31].

On the other hand, in 2013, the 'firewall' proponents Almheiri, Marolf, Polchinski and Sully had argued that if the net Hawking radiation from a BH with a traditional EH is indeed in a pure state then either the information carried by the emission is not from regions close to the EH or the EH itself manifests physically in the form of a catastrophic discontinuity for a freely falling observer [32].

By providing a critical analysis of both the 'fuzzball' as well as the 'firewall' proposals, a comprehensive review of the subject of Hawking radiation has been put forward by Suvrat Raju recently that includes his own insights pertaining to the information loss puzzle [33]. According to Raju, careful calculations point to the existence of a principle of redundancy entailing retention of the full copy of the BH interior information even outside the EH, and thereby, resolving the paradox.

In other words, the jury is still out concerning the BH information loss paradox.

3. Charged Black Holes, Monopoles and Dyons

3.1. Horizon, Ergosphere, Penrose Process and Wald's Mechanism

The exact solution corresponding to the exterior metric of a spherically symmetric but electrically charged mass distribution, with total mass M and total charge Q, were derived by Reissner in 1916 and, independently, by Nordström in 1918. Although the space-time line-element has the same form as eq. (1), but because of the contribution of the electromagnetic field to the full energy-momentum tensor, the function $f(r)$ is different from the Schwarzschild case, and is given by,

$$f(r) = \frac{R_s}{r} - \frac{GQ^2}{c^4 r^2} \tag{13}$$

so that, by setting $\frac{R_s}{r} - \frac{GQ^2}{c^4 r^2} = 0$, one obtains the EH for a Reissner-Nordström BH to be at a radius,

$$R_{RN}^{EH} = \frac{GM}{c^2} + \sqrt{\left(\frac{GM}{c^2}\right)^2 - \frac{GQ^2}{c^4}} \tag{14}$$

Since gravitation (or, equivalently space-time geometry) alone does not distinguish between $+Q$ and $-Q$ charge, it is not surprising that eqs.(13) and (14) have only quadratic dependence on the electric charge, and not a linear one.

From eq. (14), it is obvious that the EH size of a Reissner-Nordström BH with $Q \neq 0$ is always less than that of a Schwarschild BH of same mass, and that it is necessary to have the condition $GM^2 \geq Q^2$ met, for the solution to be valid (from a Newtonian point of view, the latter condition means that the net potential energy of the BH due to both gravity and Coulombic force must be negative for its stability).

The most general classical BH, with mass M, angular momentum J and electric charge Q, corresponds to the Kerr-Newman solution (see e.g., [8]), with its EH size being given by,

$$R_{KN}^{EH} = \frac{GM}{c^2} + \sqrt{\left(\frac{GM}{c^2}\right)^2 - \left(\frac{J}{Mc}\right)^2 - \frac{GQ^2}{c^4}} \tag{15}$$

and its Hawking temperature by,

$$T_H^{KN} = \frac{\hbar c}{4\pi k_B R_{KN}^{EH}} \frac{(R_{KN}^{EH})^2 - (J/Mc)^2 - \frac{GQ^2}{c^4}}{(R_{KN}^{EH})^2 + (J/Mc)^2}\,, \tag{16}$$

which reduce to eqs.(3) and (11), respectively, when $J = 0 = Q$. One recovers the corresponding Kerr BH expressions by setting $Q = 0$ in eqs.(15) and (16).

As in the case of a Kerr BH, a Kerr-Newman BH too possesses an ergosphere [34] where test particles can have negative energy (as measured by inertial observers far away). Suppose a massive test particle with an initial positive energy E enters the ergosphere, and then splits up into two particles A and B, with energy $E_A > 0$ and $E_B < 0$, respectively. Since Kerr BHs are stationary BHs, admitting a time-like Killing vector $\xi^\mu = \delta^\mu_0$, a particle's energy is conserved i.e., $E = E_A + E_B$. As $E_B < 0$, it follows that $E_A > E$. Hence, if the particle A emerges out of the ergosphere, one has mined energy out of the rotating BH at the expense of latter's rotational energy. This is the well known 'Penrose process' of energy extraction [35, 36, 37].

One may wonder how to prevent an electrically charged BH (even after creating it somehow) getting neutralized by attracting oppositely charged particles from an ambient plasma, since the former has an associated long range electric field. Interestingly enough, in 1974, Wald had provided a mechanism by which a Kerr BH, embedded in an ambient plasma as well as a uniform magnetic field, would eventually turn into a Kerr-Newman BH [38].

Using the Killing vector equations ingeniously, Wald had obtained an exact solution representing a Kerr BH of mass M and spin angular momentum $\vec{J}$ along with a magnetic field surrounding it which, at very large distances, assumed a constant value $\vec{B}_0$ (either parallel or anti-parallel to $\vec{J}$). Furthermore, he had shown that ambient charged particles of a given sign would be drawn preferentially into the BH, where the sign of the charge depended on the sign of $(\vec{B}_0.\vec{J})$.

The charge Q of such a Kerr-Newman BH would satisfy the constraint [38],

$$\frac{Q}{(GM/c^2)^2} \leq 2B_0 \quad . \tag{17}$$

Now, astrophysical BHs are all likely to be Kerr BHs, as majority of (progenitor) stars exhibit rotation. Also, massive stars tend to be found in dense, gaseous environments. In other words, since magnetic field associated with cosmic plasma is ubiquitous, there is a strong possibility of finding electrically charged BHs in the dense regions of stellar clusters, if one goes by Wald's solution.

3.2. Magnetic Monopoles and Dyons

Electrically charged BH solutions can also be generalized to those with magnetic and other gauge charges [39, 40, 42, 41]. Although no magnetic monopole (i.e., a particle with a magnetic charge) has been seen in the nature so far, the plausible existence of magnetic monopoles not only enlarges the symmetry in electrodynamics, it also leads to a charge quantization condition [43, 44]. An electromagnetic duality transformation, a rotation in an internal space that mixes up both electric and magnetic fields as well as the electric and magnetic charges in the same way, is an exact symmetry of the extended Maxwell equations.

Furthermore, with quantum theory juxtaposed, the singular nature of the vector potential $\vec{A}(\vec{r}, t)$ associated with a monopole with magnetic charge g and the requirement that the wavefunction describing a test particle with electric charge q be single valued, lead to the Dirac quantization, $qg = n\hbar c/2$, thereby making the 'magnetic fine structure constant' $\alpha_{mag} \equiv g^2/\hbar c > 34.25$ [43, 45].

The concept of a dyon (a hypothetical particle carrying both electric as well as magnetic charge, (q, g)) was put forward by Schwinger in a paper that emphasized, among other interesting ideas, on the U(1) global electromagnetic duality symmetry as well as on the ensuing large α_{mag} [46]. Attempts to gauge the U(1) electromagnetic duality symmetry have also been made in the past [47, 48, 49].

The main motivation of Schwinger's 1969 paper was to provide a model of hadronic matter being composed of dyons bearing fractional electric charges. This idea had an uncanny similarity with Saha's speculative paper of 1936 surmising that neutrons, because of their non-vanishing magnetic dipole moment, are made of monopole-anti monopole pairs [44]. In this paper, Saha also reproduced Dirac's seminal charge quantization rule by invoking quantization of the angular momentum of the electromagnetic field (see [44, 50, 51, 52, 53, 54]).

Of course, in the string theory paradigm, interesting mathematical solutions, representing dyons and dyonic BHs with characteristic spectrum, have already been obtained (see e.g., [55, 56, 57, 58, 59, 60]). Similarly, magnetic monopole as well as dyon solutions, including dyonic BHs, emerge naturally from non-Abelian gauge theories, with monopole mass as high as $\sim 10^{16}$ GeV, as in the case of GUT models [61, 62, 63, 64, 65, 66, 67, 68, 69].

What is perhaps more interesting is that because U(1) subgroup of SU(2) in the Salam-Weinberg model of electroweak unification can have non-trivial homotopy, the standard model entails presence of topologically stable magnetic monopoles weighing about 4 to 7 TeV ([65], [66], [70]). The estimated mass range makes the electroweak monopole quite appealing, as its existence can be corroborated in the near future by the MoEDAL detector at LHC ([71], [72]).

Moreover, the electroweak phase transition that occurred when the universe was $\sim 10^{-11}$ s old and was very hot, corresponding to a temperature $k_B T \sim 100$ GeV, would have led to production of magnetic monopoles with a net density of $\Omega_{0,\ mono} \approx 6.5 \times 10^{-11}$ in the current epoch [73]. However, for monopoles with mass less than about 10^{17} GeV there are strong observational constraints ([74], [75]). So far, all searches for monopoles and dyons have been illusive, yielding null results.

4. Dynamics of Dyon-Dyon Interaction

4.1. Equations of Motion and the Trajectories

As an interesting theoretical problem, we begin in this section, a study of dyon-dyon interaction. We consider two dyons D1 and D2 in an inertial frame S, having electromagnetic charges (q_1, g_1) and (q_2, g_2), masses m_1, m_2 and instantaneous position vectors $\vec{r}_1(t)$, $\vec{r}_2(t)$, respectively (see Figure 1). Apart from the electromagnetic interactions between them, their mutual gravitational attractions have also been included in the present analysis.

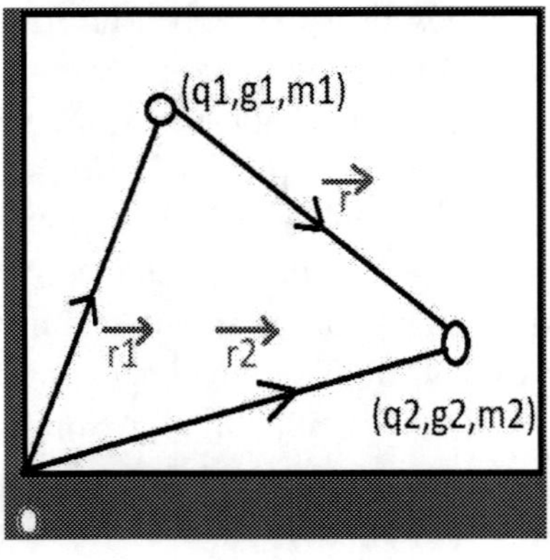

Figure 1. A pair of interacting dyons in an inertial frame S.

At an instant t and at the position vector $\vec{r}$ of the frame S, the electric and magnetic fields sourced by D1 as well as D2 in motion, are given by,

$$\vec{E}(\vec{r},t) = \vec{E}_1(\vec{r},t) + \vec{E}_2(\vec{r},t) = \left[q_1 \frac{\vec{r}-\vec{r}_1(t)}{|\vec{r}-\vec{r}_1(t)|^3} - g_1 \frac{\dot{\vec{r}}_1 \times (\vec{r}-\vec{r}_1(t))}{c|\vec{r}-\vec{r}_1(t)|^3}\right] + \left[q_2 \frac{\vec{r}-\vec{r}_2(t)}{|\vec{r}-\vec{r}_2(t)|^3} - g_2 \frac{\dot{\vec{r}}_2 \times (\vec{r}-\vec{r}_2(t))}{c|\vec{r}-\vec{r}_2(t)|^3}\right] \quad (18)$$

and,

$$\vec{B}(\vec{r},t) = \vec{B}_1(\vec{r},t) + \vec{B}_2(\vec{r},t) = \left[g_1 \frac{\vec{r}-\vec{r}_1(t)}{|\vec{r}-\vec{r}_1(t)|^3} + q_1 \frac{\dot{\vec{r}}_1 \times (\vec{r}-\vec{r}_1(t))}{c|\vec{r}-\vec{r}_1(t)|^3}\right] + \left[g_2 \frac{\vec{r}-\vec{r}_2(t)}{|\vec{r}-\vec{r}_2(t)|^3} + q_2 \frac{\dot{\vec{r}}_2 \times (\vec{r}-\vec{r}_2(t))}{c|\vec{r}-\vec{r}_2(t)|^3}\right], \quad (19)$$

respectively. In the above equations, $\vec{E}_1$ ($\vec{E}_2$) is the electric field due to D1 (D2) while $\vec{B}_1$ ($\vec{B}_2$) is the magnetic field generated by D1 (D2). (We have adopted the standard notations $\dot{f} \equiv df/dt$ and $\ddot{f} \equiv d^2f/dt^2$ throughout.)

It is fairly straightforward to write down the equations of motion that govern the classical dynamics of a pair of dyons (e.g., [76]). What is new in this article is that we have extended the analysis by incorporating the mutual gravitational attraction between the pair,

$$m_1\ddot{\vec{r}}_1 = q_1[\vec{E}_2(\vec{r}_1,t) + \frac{\dot{\vec{r}}_1}{c} \times \vec{B}_2(\vec{r}_1,t)] + g_1[\vec{B}_2(\vec{r}_1,t) - \frac{\dot{\vec{r}}_1}{c} \times \vec{E}_2(\vec{r}_1,t)] - Gm_1m_2\frac{\vec{r}_1-\vec{r}_2}{|\vec{r}_1-\vec{r}_2|^3} \quad (20)$$

$$= \frac{(-Gm_1m_2 + q_1q_2 + g_1g_2)(\vec{r}_1-\vec{r}_2)}{|\vec{r}_1-\vec{r}_2|^3} + \frac{(q_1g_2 - q_2g_1)(\dot{\vec{r}}_1 - \dot{\vec{r}}_2) \times (\vec{r}_1-\vec{r}_2)}{c|\vec{r}_1-\vec{r}_2|^3} \quad (21)$$

and,

$$m_2\ddot{\vec{r}}_2 = q_2[\vec{E}_1(\vec{r}_2,t) + \frac{\dot{\vec{r}}_2}{c} \times \vec{B}_1(\vec{r}_2,t)] + g_2[\vec{B}_1(\vec{r}_2,t) - \frac{\dot{\vec{r}}_2}{c} \times \vec{E}_1(\vec{r}_2,t)] + Gm_1m_2\frac{\vec{r}_1-\vec{r}_2}{|\vec{r}_1-\vec{r}_2|^3} \quad (22)$$

$$= \frac{(-Gm_1m_2 + q_1q_2 + g_1g_2)(\vec{r}_2 - \vec{r}_1)}{|\vec{r}_1 - \vec{r}_2|^3} + \frac{(q_2g_1 - q_1g_2)(\dot{\vec{r}}_2 - \dot{\vec{r}}_1) \times (\vec{r}_2 - \vec{r}_1)}{c|\vec{r}_1 - \vec{r}_2|^3} . \tag{23}$$

In the present study, D1 and D2 are assumed to be moving with non-relativistic speeds in S, so that $v_i \equiv |\frac{d\vec{r}_i}{dt}| \ll c$, $i = 1, 2$. Therefore, terms of the order $\mathcal{O}(v_i^2/c^2)$ that appear in the dynamical equations have been neglected. For instance, the Lorentz force on D2 due to a magnetic field $\vec{B}_{12} = \frac{q_1\dot{\vec{r}}_1 \times (\vec{r}_2 - \vec{r}_1)}{c|\vec{r}_1 - \vec{r}_2|^3}$ generated by the motion of D1 at $\vec{r}_2$, etc. have been omitted since such terms are of $\mathcal{O}(v^2/c^2)$ order. One may refer to the study by Das and Majumdar for their analysis of charge-monopole scattering at Planckian energies [77], and [78] for non-relativistic dyon-dyon quantum scattering.

The separation between D1 and D2 represented by the relative vector $\vec{r}$, the position vector of the centre of mass (CM) $\vec{R}$ and the reduced mass are defined by,

$$\vec{r}(t) = \vec{r}_2(t) - \vec{r}_1(t) \, , \;\; \vec{R}(t) = \frac{m_1\vec{r}_1(t) + m_2\vec{r}_2(t)}{m_1 + m_2} \, , \;\; \text{and} \;\; \mu = \frac{m_1m_2}{m_1 + m_2} \, , \tag{24}$$

respectively, so that eqs.(21) and (23) can be expressed in terms of μ, $\vec{r}$ and $\vec{R}$ as follows,

$$\ddot{\vec{R}} = \vec{0} \Rightarrow \dot{\vec{R}} = \vec{V}_{CM} = \text{constant vector} \, , \tag{25}$$

and,

$$\mu \frac{d^2\vec{r}}{dt^2} = -\frac{1}{r^3}(b\vec{r} + a\vec{l}), \tag{26}$$

where, $a \equiv (q_2g_1 - q_1g_2)/(\mu c)$, $b \equiv Gm_1m_2 - (q_1q_2 + g_1g_2)$ and $\vec{l} = \mu\vec{r} \times \dot{\vec{r}}$ is the angular momentum of the pair in the CM-frame.

It is easy to derive three constants of motion ensuing from eq. (26),

$$E = \frac{1}{2}\,\mu\,\dot{r}^{\,2} + \frac{l^2}{2\mu r^2} - \frac{b}{r}, \tag{27}$$

$$\vec{L} = \vec{l} - \frac{a\mu\vec{r}}{r} \tag{28}$$

$$\text{and } l^2 = L^2 - \mu^2a^2 \, , \tag{29}$$

where E and $\vec{l}$ are the energy and the angular momentum, respectively, of the reduced system in the CM frame. In the CM-frame, $\vec{V}_{CM}$ is, of course, identically zero.

Making use of eq. (28), eq. (26) may be re-expressed as,

$$\frac{d^2\vec{r}}{dt^2} = -\frac{1}{\mu\, r^3}\left[\left(b + \frac{\mu a^2}{r}\right)\vec{r} + a\vec{L}\right] . \tag{30}$$

Since $\vec{L}$ remains constant, we may take the z-axis to be along this constant vector. Then, adopting spherical polar coordinates and making use of eqs.(28) and (29), we have with us,

$$\vec{r}.\vec{L} = rL\cos\theta_0 = -a\mu\, r \Rightarrow \cos\theta_0 = -\frac{a\mu}{L} \text{ and } \sin\theta_0 = \pm\frac{l}{L} , \tag{31}$$

implying that the vector $\vec{r}(t)$, representing the separation between the two dyons, always lies on a cone of half-opening angle θ_0 with its apex at the origin.

From pure kinematics, i.e., by expanding $\vec{l}\times\vec{r}$, one can also derive an useful expression,

$$\mu\dot{\vec{r}} = \frac{1}{r}\left(\mu\,\dot{r}\,\vec{r} - \frac{\vec{r}}{r}\times\vec{l}\right) = \frac{1}{r}\left(\mu\,\dot{r}\,\vec{r} - \frac{\vec{r}}{r}\times\vec{L}\right) \tag{32}$$

so that, with $\vec{L} = L\hat{e}_z$, we get an expected result,

$$\dot{z} = \frac{\dot{r}}{r}z \quad \Rightarrow \quad z(t) = r(t)\cos\theta_0 \tag{33}$$

Moreover, with this choice of the z-axis, we obtain from eq. (30),

$$\ddot{x} = -\frac{1}{\mu\, r^3}\left[b + \frac{\mu a^2}{r}\right]x , \tag{34}$$

and,

$$\ddot{y} = -\frac{1}{\mu\, r^3}\left[b + \frac{\mu a^2}{r}\right]y , \tag{35}$$

so that, $x = r\sin\theta_0\cos\phi$ and $y = r\sin\theta_0\sin\phi$ leads to the instantaneous angular speed,

$$\omega(t) = \dot{\phi} = \pm\frac{L}{\mu\, r^2} . \tag{36}$$

Above considerations allow one to immediately obtain circular orbit solutions ($\dot{r} = 0$ and $r(t) = r_0$ = constant) of eq. (26). In this case, it is trivial to see from eq. (33) that $z_0 = -a\mu r_0/L$ constitutes the orbital plane. The radius of the circular orbit is given by $\rho_0 = \sqrt{r_0^2 - z_0^2} = lr_0/L$. Employing eqs.(27) and (36), one obtains for the orbit,

$$r_0 = \begin{cases} \frac{l^2}{2\mu b} \Rightarrow b > 0, & E = 0 \\ \frac{l^2}{\mu b}\left[1 \pm \sqrt{1 + 2l^2 E/\mu b^2}\right]^{-1}, & E \neq 0 \end{cases} \tag{37}$$

and a constant angular speed,

$$\omega = \pm \frac{L}{\mu\, r_0^2} \tag{38}$$

4.2. Charge Quantization Condition for Dyons

Saha's argument for charge quantization can be extended trivially to the case of two static dyons in order to arrive at the Schwinger-Zwanziger quantization condition [79, 76, 80]. Referring to Figure 1 as well as eqs.(18) and (19), if one considers the dyons to be at rest (i.e., $\dot{\vec{r}}_1 = 0 = \dot{\vec{r}}_2$), the net electric and magnetic fields at $\vec{r}_P$ due to D1 and D2 are given by,

$$\overrightarrow{E}(\overrightarrow{r}_P, t) = q_1 \frac{(\overrightarrow{r}_{P-r_1(\overrightarrow{t})})}{|\overrightarrow{r}_P - \overrightarrow{r}_1(t)|^3} + q_2 \frac{(\overrightarrow{r}_P - \overrightarrow{r}_2(t))}{|\overrightarrow{r}_P - \overrightarrow{r}_2(t)|^3} \tag{39}$$

and,

$$\vec{B}(\vec{r}_P, t) = g_1 \frac{(\vec{r}_P - \vec{r}_1(t))}{|\vec{r}_P - \vec{r}_1(t)|^3} + g_2 \frac{(\vec{r}_P - \vec{r}_2(t))}{|\vec{r}_P - \vec{r}_2(t)|^3} \tag{40}$$

so that the Poynting vector is proportional to,

$$\vec{E}(\vec{r}_P, t) \times \vec{B}(\vec{r}_P, t) = (q_1 g_2 - q_2 g_1) \frac{(\vec{r}_P - \vec{r}_1(t)) \times (\vec{r}_P - \vec{r}_2(t))}{|\vec{r}_P - \vec{r}_1(t)|^3 |\vec{r}_P - \vec{r}_2(t)|^3} \tag{41}$$

The EM field angular momentum due to two static dyons then is simply,

$$\vec{l}_{EM}(t) = \frac{1}{4\pi c} \int \vec{r}_P \times (\vec{E}(\vec{r}_P, t) \times \vec{B}(\vec{r}_P, t)) d^3 r_P \tag{42}$$

After integrating over $\vec{r}_P$, one obtains,

$$\vec{l}_{EM}(t) = -\frac{q_1 g_2 - q_2 g_1}{c} \frac{\vec{r}_2(t) - \vec{r}_1(t)}{|\vec{r}_2(t) - \vec{r}_1(t)|} = -\frac{q_1 g_2 - q_2 g_1}{c} \frac{\vec{r}(t)}{r(t)} \tag{43}$$

(The above integration can be performed trivially by assuming D1 to be at the origin of the coordinate system, without any loss of generality, so that $\vec{r_1} = 0$.)

Applying now Saha's argument that angular momentum of the electromagnetic field must be quantized as $|\vec{l}_{EM}| = n\hbar/2$, one reaches the desired result,

$$\frac{q_1 g_2 - q_2 g_1}{c} = n\hbar/2\,, \tag{44}$$

namely, the well known Schwinger-Zwanziger quantization condition.

4.3. Scalar Virial Theorem for a Self-Gravitating System Containing Dyons

Suppose one is interested in astrophysical BHs acquiring both electric and magnetic charges by accreting dyons statistically from the environment. For such a possibility, the stellar mass BHs must belong to a large virialized system since the time scale on which the BHs can become charged, in a stochastic manner, is expected to be of the order of billion years. Hence, it is instructive to take a closer look at the virial equilibrium condition for a self-gravitating system consisting of baryonic matter, dark matter, stellar mass BHs as well as dyons. For simplicity, we adopt a point particle description for each of the constituent of such a astrophysical cloud.

We begin our study with a large cosmic nebula consisting of N randomly moving particles/dyons characterized by their position vectors $\vec{r}_i(t)$, masses m_i and electromagnetic charges (q_i, g_i), $i = 1, 2, ..., N$. For the ones that are electrically neutral, $q_j = 0 = g_j$. Eqs.(21) and (23) can be extended in a straightforward manner to express the force experienced by the jth particle/dyon due to rest of the particles,

$$\vec{F}_j = m_i \ddot{\vec{r}}_j = \sum_{k=1,k\neq j}^{N} \frac{1}{|\vec{r}_j - \vec{r}_k|^3} \Big[(-Gm_j m_k + q_j q_k + g_j g_k)(\vec{r}_j - \vec{r}_k)$$

$$\frac{q_j q_k - q_j q_k}{c}(\dot{\vec{r}}_j - \dot{\vec{r}}_k) \times (\vec{r}_j - \vec{r}_k)\Big] \tag{45}$$

Taking the dot product of eq. (45) with $\dot{\vec{r}}_j$ and then summing over all the particles, we get,

$$\sum_{j=1}^{N}\sum_{k=1,k\neq j}^{N} \dot{\vec{r}}_j.\vec{F}_j = \frac{d}{dt}\sum_{j=1}^{N}\sum_{k=1,k\neq j}^{N} \frac{1}{2}m_i\dot{\vec{r}}_j.\dot{\vec{r}}_j \equiv \frac{dT}{dt} = \tag{46}$$

$$= \sum_{j=1}^{N}\sum_{k=1,k\neq j}^{N}$$

$$\frac{\dot{\vec{r}}_j \,.}{|\vec{r}_j - \vec{r}_k|^3}\left[(-Gm_jm_k + q_jq_k + g_jg_k)(\vec{r}_j - \vec{r}_k) + \frac{q_jg_k - q_jg_k}{c}(\dot{\vec{r}}_j - \dot{\vec{r}}_k) \times (\vec{r}_j - \vec{r}_k)\right] \tag{47}$$

Considering half of the RHS of eq. (47) with indices j and k interchanged as well as using the anti-symmetry of $q_jg_k - q_jg_k$ in j and k, we may express the rate of change of the kinetic energy of the system as,

$$\frac{dT}{dt} = \frac{1}{2}\sum_{j=1}^{N}\sum_{k=1,k\neq j}^{N} \frac{(\dot{\vec{r}}_j - \dot{\vec{r}}_k)\,.}{|\vec{r}_j - \vec{r}_k|^3}\left[(-Gm_jm_k + q_jq_k + g_jg_k)(\vec{r}_j - \vec{r}_k) + \frac{q_jg_k - q_jg_k}{c}(\dot{\vec{r}}_j - \dot{\vec{r}}_k) \times (\vec{r}_j - \vec{r}_k)\right] \tag{48}$$

The dot product of $\dot{\vec{r}}_j - \dot{\vec{r}}_k$ with the second term of the above equation vanishes identically, reflecting the fact that work done by magnetic (electric) fields on electric (magnetic) charges is zero.

For any vector $\vec{a}$, we have,

$$\frac{d}{dt}\left(\frac{1}{|\vec{a}|}\right) = -\frac{\vec{a}.\dot{\vec{a}}}{|\vec{a}|^3}\,,$$

and hence, eq. (48) can be expressed as,

$$\frac{dT}{dt} = -\frac{d}{dt}\left(\frac{1}{2}\sum_{j=1}^{N}\sum_{k=1,k\neq j}^{N} \frac{-Gm_jm_k + q_jq_k + g_jg_k}{|\vec{r}_j - \vec{r}_k|}\right) = -\frac{dV}{dt} \tag{49}$$

$\Rightarrow E \equiv T + V$ is the conserved energy, where,

$$V \equiv \frac{1}{2}\sum_{j=1}^{N}\sum_{k=1,k\neq j}^{N} \frac{-Gm_jm_k + q_jq_k + g_jg_k}{|\vec{r}_j - \vec{r}_k|} \tag{50}$$

is the potential energy of the cloud.

Now, the centre of mass position vector is simply, $\vec{R} \equiv \frac{\sum_i^N m_i\vec{r}_i}{\sum m_i} = \frac{\sum_i^N m_i\vec{r}_i}{M}$, where $M = \sum_i^N m_i$ is the total mass of the cosmic cloud. Using eq. (45) one can easily demonstrate that,

$$\ddot{\vec{R}} = \frac{1}{M}\sum m_j\ddot{\vec{r}}_j \tag{51}$$

$$= \frac{1}{M}\sum_{j=1}^{N}\sum_{k=1,k\neq j}^{N} \frac{1}{|\vec{r}_j - \vec{r}_k|^3}\Bigg[(-Gm_jm_k + q_jq_k + g_jg_k)(\vec{r}_j - \vec{r}_k) + \frac{q_jg_k - q_jg_k}{c}(\dot{\vec{r}}_j - \dot{\vec{r}}_k) \times (\vec{r}_j - \vec{r}_k)\Bigg] = 0 \tag{52}$$

because of the anti-symmetry of the force term that is being summed over with respect to the indices j and k. Therefore, the kinetic energy associated with centre of mass motion, $T_{CM} = \frac{1}{2}M\dot{R}^2$ is a constant of motion so that, from eqs.(49) and (50), we have,

$$E - T_{CM} = T - T_{CM} + V = \sum_{i=1}^{N}\frac{1}{2}m_i\dot{\vec{r}}_i.\dot{\vec{r}}_i - \frac{1}{2}M\dot{\vec{R}}.\dot{\vec{R}} + V =$$

$$= \frac{1}{2M}\sum_{i=1}^{N}\sum_{j=1}^{N} m_im_j(\dot{\vec{r}}_i.\dot{\vec{r}}_i - \dot{\vec{r}}_i.\dot{\vec{r}}_j) + V \tag{53}$$

$$= \frac{1}{2M}\sum_{i=1}^{N}\sum_{j=1}^{N} m_im_j\dot{\vec{r}}_i.(\dot{\vec{r}}_i - \dot{\vec{r}}_j) + V =$$

$$\frac{1}{4M}\sum_{i=1}^{N}\sum_{j=1}^{N} m_im_j(\dot{\vec{r}}_i - \dot{\vec{r}}_j).(\dot{\vec{r}}_i - \dot{\vec{r}}_j) + V \tag{54}$$

to be also a constant of motion. In particular, if $T - T_{CM} + V < 0$, the self-gravitating cloud would be a bound system. Of course, this necessitates $V < 0$.

One may also obtain a scalar virial theorem for such a massive object made of dyon-particle-BH mixture. Given the moment of inertia, $I = \sum_{i=1}^N m_i\vec{r}_i.\vec{r}_i$, equations of motion entail [81],

$$\frac{1}{2}\frac{d^2I}{dt^2} = 2T + \sum_{j=1}^{N}\vec{r}_j.\vec{F}_j \ . \tag{55}$$

In our case, making use of eq. (45), the second term of the RHS of eq. (55) reduces to,

$$\sum_{j=1}^{N} \vec{r}_j.\vec{F}_j = \sum_{j=1}^{N} \sum_{k=1,k\neq j}^{N} \frac{\vec{r}_j}{|\vec{r}_j - \vec{r}_k|^3}. \Big[(-Gm_j m_k + q_j q_k + g_j g_k)(\vec{r}_j - \vec{r}_k) + \frac{q_j g_k - q_j g_k}{c}(\dot{\vec{r}}_j - \dot{\vec{r}}_k) \times (\vec{r}_j - \vec{r}_k)\Big] \quad (56)$$

$$= \frac{1}{2}\sum_{j=1}^{N} \sum_{k=1,k\neq j}^{N} \frac{(\vec{r}_j - \vec{r}_k).}{|\vec{r}_j - \vec{r}_k|^3} \Big[(-Gm_j m_k + q_j q_k + g_j g_k)(\vec{r}_j - \vec{r}_k) + \frac{q_j g_k - q_j g_k}{c}(\dot{\vec{r}}_j - \dot{\vec{r}}_k) \times (\vec{r}_j - \vec{r}_k)\Big] = V \ , \quad (57)$$

where V is given by eq. (50).

If we define,

$$I_0 \equiv I - M\vec{R}.\vec{R} \quad (58)$$

then, using eqs.(52), (53), (55) and (57), it can be easily shown that the scalar virial theorem takes the form,

$$\frac{1}{2}\ddot{I}_0 = \frac{1}{2}\ddot{I} = 2T + V - M\dot{\vec{R}}.\dot{\vec{R}} = 2(T - T_{CM}) + V \quad (59)$$

In the continuum limit, the potential energy,

$$\sum_{i=1}^{N} \sum_{j=1,i\neq j}^{N} \frac{-Gm_i m_j + q_i q_j + g_i g_j}{|\vec{r}_i - \vec{r}_j|} \quad (60)$$

becomes,

$$\rightarrow V = \int d^3r \int d^3r' \, \frac{-G\rho(\vec{r},t)\rho(\vec{r'},t) + \rho_{el}(\vec{r},t)\rho_{el}(\vec{r'},t) + \rho_{mag}(\vec{r},t)\rho_{mag}(\vec{r'},t)}{|\vec{r} - \vec{r'}|} \quad (61)$$

where ρ, ρ_{el} and ρ_{mag} are the mass, electric and magnetic charge densities, respectively.

For a near spherical, uniformly dense cloud of radius $\sim R$, mass M, net electric charge Q_e and net magnetic charge Q_m, the above expression leads to,

$$V \sim \frac{-GM^2 + Q_e^2 + Q_m^2}{R} \quad (62)$$

so that, using eqs.(59) and (62), the conditions for the cosmic ball to be in a bound state as well as in virial equilibrium (i.e., $\ddot{I}_0 = 0$) take the form,

$$Q_e^2 + Q_m^2 < GM^2 \text{ and } < T > - T_{CM} \cong -\frac{1}{2} < V > \quad . \tag{63}$$

As long as the conditions given by eq. (63) are met, a cosmic cloud containing a large number of dyons, among other entities like stars and BHs, can certainly be in a state of virial equilibrium. BHs (initially neutral), lying in such astrophysical clouds, can absorb dyons statistically and become electromagnetically charged gradually. For instance, if a BH randomly accretes N_{dyon} number of dyons, its electromagnetic charge will eventually be proportional to $\sqrt{N_{\text{dyon}}}$ (see e.g., [40], and the references therein).

5. Radiation from Interacting Dyons: Electromagnetic and Gravitational Waves

In any theory of fields that is consistent with special relativity (SR), the field perturbations would propagate as waves with speeds $v \leq c$. This is because, all physical perturbations must bear energy, since only those entities that carry energy can be detected or measured, for the simple reason that exchange of energy between an entity and a sensing device is necessary for the former's detection. Even in quantum physics this is true since a quantum system and a measurement apparatus can influence each other only through a suitable interaction Hamiltonian. Now, an energy carrying perturbation has an associated mass via $E = mc^2$ relation. Hence, because of the result $m = m_0/\sqrt{1 - v^2/c^2}$ of SR, m_0 being the rest mass, the propagation speed of any measurable perturbation being $\leq c$ is a stringent requirement in any physical theory.

Both electrodynamics as well as GR are consistent with SR. Thus, it ensues that electromagnetic (EM) waves and gravitational radiation (i.e., undulations in space-time geometry) travel in vacuum with speed c. Existence of gravitational waves (GWs) predicted by GR was indirectly corroborated by the meticulous observations of the binary pulsar system PSR 1913 + 16 by Taylor and Hulse, who demonstrated that the orbital period of PSR 1913 + 16 decreased in accordance with the loss of orbital energy due to the emission of GWs ([82], and the references therein). The first ever direct detection of GWs from a binary black hole system involving 36 $M_\odot$ and 29 $M_\odot$ BHs was achieved by LIGO in 2015 [83].

Emission of EM waves and GWs from pure electrically charged BH binary was studied by Leibling and Palenzuela in 2016 [84]. The primary objective of this section is to consider such radiation from a pair of interacting dyonic BHs, moving non-relativistically. Since we are interested in studying orbiting heavy dyon pairs of astrophysical significance, the precise nature of their orbital trajectories needs to be studied. From eq. (27), the rate at which the distance between the pair changes is given by,

$$\dot{r} = \pm\sqrt{\frac{2}{\mu}\Big(E + \frac{b}{r} - \frac{l^2}{2\mu r^2}\Big)} \tag{64}$$

The above equation can easily be integrated to yield,

$$t = K \pm \frac{1}{E}\sqrt{\frac{\mu}{2}}\Big[\sqrt{Er^2 + br - \frac{l^2}{2\mu}} - \frac{b}{2}I_1\Big] \tag{65}$$

where,

$$I_1 \equiv \begin{cases} \frac{1}{\sqrt{E}}\ln\Big(2\sqrt{E(Er^2+br-\frac{l^2}{2\mu})} + 2Er + b\Big), & E>0 \\ \frac{1}{\sqrt{E}}\sinh^{-1}\Big(\frac{2Er+b}{\sqrt{\Delta}}\Big), & E>0, \Delta>0 \\ -\frac{1}{\sqrt{-E}}\sin^{-1}\Big(\frac{2Er+b}{\sqrt{-\Delta}}\Big), & E<0, \Delta<0 \\ \frac{1}{\sqrt{E}}\ln\Big(2Er+b\Big), & E>0, \Delta=0 \end{cases} \tag{66}$$

and $-\Delta \equiv b^2 + \frac{2l^2E}{\mu}$, with K being a constant of integration.

Expressing eq. (28) as,

$$\mu\vec{r}\times\dot{\vec{r}} = \vec{L} + \frac{a\mu\vec{r}}{r} \quad , \tag{67}$$

one can obtain the following equivalent result by taking a cross product of the above equation with $\vec{r}$,

$$\mu\dot{\vec{r}} = \frac{1}{r}\Big(\mu\dot{r}\vec{r} - \frac{\vec{r}}{r}\times\vec{L}\Big) \tag{68}$$

so that one may use eqs.(64) and the constancy of $\vec{L}$ in the above to solve for $\vec{r}(t)$.

Choosing the z-axis to be in the direction of $\vec{L}$ (hereafter, we will assume that $\vec{L} = L\hat{z}$ throughout the paper) and making use of eq. (65), we can integrate eq. (68) to obtain,

$$x(t) = r(t)\sqrt{1-\alpha^2}\cos(\phi(t)+K_1), y(t) = r(t)\sqrt{1-\alpha^2}\sin(\phi(t)+K_1), z(t) = \\ = \alpha\, r(t) \quad (69)$$

$$\alpha \equiv -a\mu/L,\ \phi(t) \equiv \pm\frac{L}{\sqrt{2\mu}}\begin{cases} \frac{\sqrt{2\mu}}{l}\sin^{-1}\left(\frac{br-l^2/\mu}{r\sqrt{-\Delta}}\right), & \Delta < 0 \\ \frac{\sqrt{2\mu}}{l}\tan^{-1}\left(\frac{\sqrt{\mu/2}(br-l^2/\mu)}{l\sqrt{Er^2+br-\frac{l^2}{2\mu}}}\right), & \\ -\frac{2\sqrt{br+Er^2}}{br}, & l=0 \end{cases} \quad (70)$$

with K_1 being a constant of integration and $r(t)$ obtained from eq. (65). The instantaneous angular frequency is given by,

$$\omega(t) = \frac{d\phi}{dt} = \frac{L}{\mu r^2(t)} \quad (71)$$

Because of the non-central nature of the net force mediated between the pair, the motion is confined to the surface of a cone (eqs.(31), (69) and (70)) instead of a plane. If the pair is bound, i.e., $E < 0$, one can easily demonstrate from eqs.(65) and (66) that there are two turning points corresponding to r_1 and r_2, representing the minimum and maximum separation between the pairs, respectively.

As long as the closest distance of approach between the dyonic BHs is much larger than a scale determined by their gravitational radii i.e., $r_1 > 10\ R_{BH} = 2Gm_2/c^2$, m_2 being the larger of the two BH masses, their speeds would remain non-relativistic and, hence, would be amenable to the above analysis.

In what follows, we take up the case of EM waves and GWs emission by a pair of dyons, and obtain the formal expressions for amplitudes and powers that make use of the above equations. Thereafter, by considering various parameter sets, we tabulate the estimated quantities numerically from the expressions derived theoretically, in the following subsections.

5.1. Electromagnetic Dipole Radiation

Since from eqs. (65), (66), (69) and (70), the trajectories of D1 and D2 have been determined analytically, it is quite straightforward to calculate the EM wave amplitude and the power resulting from the interacting pair.

Given the electric and magnetic dipole moments, $\vec{d} = q_1\vec{r}_1 + q_2\vec{r}_2$ and $\vec{m} = g_1\vec{r}_1 + g_2\vec{r}_2$, respectively, one can show using eqs.(21) and (23) that,

$$\ddot{\vec{d}} = -\frac{m_1 q_2 - m_2 q_1}{m_1 m_2 r^3}\left[b\vec{r} + \left(\frac{\mu a^2 \vec{r}}{r} + a\vec{L}\right)\right] \tag{72}$$

and,

$$\ddot{\vec{m}} = -\frac{m_1 g_2 - m_2 g_1}{m_1 m_2 r^3}\left[b\vec{r} + \left(\frac{\mu a^2 \vec{r}}{r} + a\vec{L}\right)\right], \tag{73}$$

so that the retarded electric and magnetic field amplitudes at the point of observation $\vec{r}_{\text{obs}}$, very far away from the pair of dyons, are given by ([85]),

$$\vec{E}(\vec{r}_{\text{obs}}, t) = \frac{1}{c^2 r_{\text{obs}}}[(\ddot{\vec{d}} \times \hat{n}) \times \hat{n} + \hat{n} \times \ddot{\vec{m}}]_{t-r_{\text{obs}}/c} \tag{74}$$

and

$$\vec{B}(\vec{r}_{\text{obs}}, t) = \frac{1}{c^2 r_{\text{obs}}}[(\ddot{\vec{m}} \times \hat{n}) \times \hat{n} - \hat{n} \times \ddot{\vec{d}}]_{t-r_{\text{obs}}/c} \,. \tag{75}$$

Making use of eqs.(72) and (73) in eqs.(74) and (75), one obtains for the electromagnetic wave amplitude the following expressions,

$$\vec{E}(\vec{r}_{\text{obs}}, t) = \frac{1}{r_{\text{obs}} c^2}\left[\left(\frac{b + \frac{\mu a^2}{r}}{r^3}\right)\Big((Bx - Ay)\hat{y} - (Ax + By)\hat{x}\Big)\right]_{t-r_{\text{obs}}/c} \tag{76}$$

$$\vec{B}(\vec{r}_{\text{obs}}, t) = \hat{n} \times \vec{E}(\vec{r}_{\text{obs}}, t) \tag{77}$$

where $A = q_1/m_1 - q_2/m_2$, $B = g_1/m_1 - g_2/m_2$, and that the z-axis has been chosen along the conserved vector $\vec{L}$.

The instantaneous power radiated in the form of electromagnetic waves is given by ([85]),

$$P_{em} = \frac{2}{3c^3}(\ddot{\vec{d}}.\ddot{\vec{d}} + \ddot{\vec{m}}.\ddot{\vec{m}}) \tag{78}$$

which in our case, from eqs. (72) and (73) along with $\vec{L} = L\hat{z}$, reduces to,

$$P_{em} = \frac{2}{3c^3 r^6}(A^2 + B^2)(a^2 l^2 + b^2 r^2) \tag{79}$$

We will use the expressions given by eqs. (76), (77) and (79) to estimate the EM fields and luminosities in section VI after we choose a set of parameters in order to tabulate and discuss our results.

5.2. Gravitational Radiation

At very large distances from a source for which energy-momentum distribution varies in a non-relativistic manner, the gravitational perturbation represented by $h_{\mu\nu}$ is expected to be weak and thus one may choose a quasi-Minkowskian coordinate system to express the metric tensor as,

$$g_{\mu\nu} \approx \eta_{\mu\nu} + h_{\mu\nu}(\vec{r}, t) \, , \tag{80}$$

with the perturbation amplitude satisfying,

$$|h_{\mu\nu}(\vec{r}, t)| \ll 1 \, . \tag{81}$$

As general covariance is built into GR, out of the 10 components of $h_{\mu\nu}$ only 2 propagating modes are physical since 6 of them can be made to vanish by making suitable general coordinate transformations [86, 87]. It is these two propagating modes that correspond to a GW amplitude.

The standard practice is to adopt a harmonic coordinate condition or the transverse, traceless (TT) gauge in which the GW amplitude, made up of a linear combination of only the space-space components of $h_{\mu\nu}$ representing two distinct polarizations h_+ and $h_\times$, is normal to the direction of propagation. For instance, if a GW is propagating in the z-direction, it may have only $h_{11}^{TT} = -h_{22}^{TT} = h_+$ and $h_{12}^{TT} = h_{21}^{TT} = h_\times$ as the non-zero components. Clearly, the GW amplitude then is manifestly traceless and transverse to the z-direction.

The GW amplitude, in the TT-gauge, as measured by an observer at time t and at a distance r_{obs} from the source, with $r_{\text{obs}} \gg$ size of the source, is given by,

$$h_{ij}^{TT}(t, \vec{r}_{\text{obs}}) = \frac{2G}{c^4 r_{\text{obs}}} \frac{d^2 Q_{ij}^{TT}(t - r_{\text{obs}}/c)}{dt^2} \, , \tag{82}$$

where Q_{ij}^{TT} is obtained from the mass quadrupole moment of a source with mass density $\rho(t, \vec{r})$,

$$Q^{ij}(t) \equiv \int \rho(t, \vec{r}) x^i x^j d^3 r \tag{83}$$

by applying suitable projection operators on eq. (83) [86].

The energy-momentum pseudo-tensor corresponding to GWs is given by [85],

$$T_{\mu\nu}^{GW} = \frac{c^4}{32\pi G} \left\langle h_{jk,\mu}^{TT} h_{,\nu}^{TT\, jk} \right\rangle$$

where h_{jk}^{TT} is the GW amplitude in the TT-gauge with $< ... >$ representing an average over many wavelengths (raising and lowering of indices of GW amplitude are done using the Minkowski metric tensor $\eta_{\mu\nu}$.).

The basic hurdle in obtaining a 'proper' energy-momentum tensor for the gravitational degrees of freedom is due to the fact that the metric tensor $g_{\mu\nu} \to \eta_{\mu\nu}$ and the Christoffel symbol $\Gamma^{\mu}_{\alpha\beta} \to 0$ in a local inertial frame (i.e., in a freely falling frame), S, so that any second rank tensor built using $g_{\mu\nu}$ and $\Gamma^{\mu}_{\alpha\beta}$ will vanish at an event point $\mathcal{E}$ in S. But this point $\mathcal{E}$ is arbitrary as one is free to choose a local inertial frame anywhere in the whole of the space-time manifold. This implies that such a tensor is identically zero every where. Using first derivatives of $\Gamma^{\mu}_{\alpha\beta}$, symmetric tensors like $R_{\mu\nu}$ or $G_{\mu\nu}$ can definitely be constructed. However, $R_{\mu\nu}$ and $G_{\mu\nu}$ vanish where there is no matter because of the Einstein equations, and thus, cannot represent energy and momentum flux of GWs traversing through vacuum. Hence, one falls back on the Landau-Lifshitz energy-momentum pseudo-tensor to study the energy and momentum associated with the gravitational degrees of freedom [85].

As T_{00}^{GW} represents the GW energy density, the GW energy flux is given by,

$$F_{gw} = cT_{00}^{GW} = \frac{c^5}{32\pi G}\left\langle h_{jk,0}^{TT} h_{,0}^{TT\ jk}\right\rangle = \frac{c^3}{32\pi G}\left\langle \dot{h}_{jk}^{TT}\dot{h}^{TT\ jk}\right\rangle$$

where $\dot{h}_{jk} \equiv \frac{\partial h_{jk}}{\partial t}$. By using eq. (82) in the above equation one obtains,

$$F_{gw} = \frac{1}{4\pi r_{\text{obs}}^2}\left(\frac{G}{2c^5}\left\langle \dddot{Q}_{jk}^{TT}\dddot{Q}^{TT\ jk}\right\rangle\right) . \tag{84}$$

Now, the surface area of a sphere of radius r is $\cong 4\pi r^2$ if the background space-time is nearly flat. So, when the emission of GWs from a source is nearly isotropic, eqs. (80), (81) and (84) entail that the GW power be given by,

$$P_{gw} \cong 4\pi r_{\text{obs}}^2\ F_{gw} = \frac{G}{2c^5}\left\langle \dddot{Q}_{jk}^{TT}\dddot{Q}^{TT\ jk}\right\rangle . \tag{85}$$

For the pair of dyonic BHs, the mass quadrupole moment Q^{ij} is given by,

$$Q^{ij} = m_1 x_1^i x_1^j + m_2 x_2^i x_2^j , \tag{86}$$

so that when one uses the relative separation vector $\vec{r}$ and the position vector $\vec{R}$ of the CM as given by eq. (24), the above equation takes the form,

$$Q^{ij} = (m_1 + m_2)X^i X^j + \mu x^i x^j . \tag{87}$$

Hence, the time derivative of the mass quadrupole moment is simply,

$$\dot{Q}^{ij} = (m_1 + m_2)[\dot{X}^i X^j + X^i \dot{X}^j] + \mu[\dot{x}^i x^j + x^i \dot{x}^j] \,. \tag{88}$$

From eq. (25), we had earlier concluded that the velocity of the CM of the pair is a constant. Hence, if we choose an inertial frame S' that is comoving with the CM, so that $\vec{R} = \vec{0}$ and $\vec{V}_{CM} = \vec{0}$, the expressions for Q^{ij} and $\dot{Q}^{ij}$ (eqs. (87) and (88)) get simplified considerably to,

$$Q^{ij} = \mu x^i x^j \,, \tag{89}$$

and,

$$\dot{Q}^{ij} = \mu[\dot{x}^i x^j + x^i \dot{x}^j] \,. \tag{90}$$

In the CM-frame, therefore, the higher time derivatives of the mass quadruple moment tensor, after making use of eq. (26), are given by,

$$\ddot{Q}_{ij} = -\frac{1}{r^3}\left[a(l_i x_j + l_j x_i) + 2b x_i x_j\right] + 2\mu \dot{x}_i \dot{x}_j \,, \tag{91}$$

and,

$$\dddot{Q}_{ij} = \frac{3\dot{r}}{r^4}\left[a(l_i x_j + l_j x_i) + \frac{2}{3} x_i x_j \left(3b + \frac{\mu a^2}{r}\right)\right] - \\ - \frac{1}{r^3}\left(4b + \frac{\mu a^2}{r}\right)\left(\dot{x}_i x_j + \dot{x}_j x_i\right) - \frac{3a}{r^3}\left(l_i \dot{x}_j + l_j \dot{x}_i\right) . \tag{92}$$

Suppose we choose the z-axis of the frame S' to be along $\vec{L}$ (note that $\vec{L}$ is a constant of motion (eq. (28)). Then, for the GW propagating along the z-axis the TT-gauge gravitational wave amplitudes corresponding to the two distinct polarizations are given by,

$$h_+(t, \vec{r}_{\text{obs}}) = \frac{1}{r_{\text{obs}}} \frac{G}{c^4} (\ddot{Q}_{xx} - \ddot{Q}_{yy})(t - r_{\text{obs}}/c) \,, \tag{93}$$

$$h_\times(t, \vec{r}_{\text{obs}}) = \frac{2G}{r_{\text{obs}} c^4} \ddot{Q}_{xy}(t - r_{\text{obs}}/c) \,, \tag{94}$$

where $\ddot{Q}_{xx}$, $\ddot{Q}_{yy}$ and $\ddot{Q}_{xy}$ are to be calculated using eq. (91) along with eqs. (65), (66), (69) and (70).

The gravitational wave power ensuing from eq. (85) has the form [86, 87, 88],

$$P_{gw} = \frac{G}{5c^5}\left(\sum_{i=1}^{2}\sum_{j=1}^{2}(\dddot{Q}_{ij}\dddot{Q}_{ij} - \frac{1}{3}\dddot{Q}_{ii}\dddot{Q}_{jj})\right),$$

so that

$$P_{gw} = \frac{G}{5c^5}\left[\frac{b^2}{r^4}\left(\frac{8}{3}\dot{r}^2 + 32\frac{l^2}{\mu^2 r^2}\right) + \frac{2a^2 l^2}{r^6}\left\{\frac{a^2}{r^2} + \frac{9l^2}{\mu^2 r^2} + \frac{8b}{\mu r}\right\}\right], \qquad (95)$$

after making use of eq. (92).

When we set $(q_i, g_i) = 0$, $i = 1, 2$, the expressions given by eqs. (91)-(95) reduce to those derived by Peter and Mathews [88], in their seminal paper describing the generation of gravitational radiation from a pair of bound uncharged mass points.

In the next section, we compute the values of GW amplitudes as well as the corresponding power for various sets of parameters.

6. Results and Discussion

If dyons that are predicted from several high energy physics theories, do exist in our universe, it is extremely crucial that measurable effects resulting from dyon-dyon interactions are studied quantitatively so that they can be verified in the near future using gravitational wave detectors, radio-telescopes and other astronomical techniques. Therefore, in this section, we consider a range of values for the parameters that determine the amplitudes and luminosities pertaining to the electromagnetic as well as gravitational radiation from a pair of bound dyons, D1 and D2. For simplicity, in our computations, we have assumed D1 to carry only magnetic charge and D2 to bear electric charge alone. In any case, because of the electromagnetic duality symmetry of the extended Maxwell equations, one can always choose a duality rotation angle such that a given dyon carries only an electric charge (or, for that matter, only a magnetic charge).

Since, in this study, we have restricted ourselves to non-relativistic dynamics for the pair D1 and D2, we have considered only those cases in which the closest distance of approach $r_1 > 18\ R_{BH}$, where $R_{BH} \equiv 2Gm_2/c^2$ with $m_2 > m_1$, and an initial relative speed $v = 0.01\ c$ for calculating the value of the angular momentum l in the CM-frame. As the BHs, in our case, are charged, the corresponding EH radii are smaller than R_{BH} and, hence, the condition

$r_1 > 18\, R_{BH}$ ensures that the Newtonian approximations are valid through out the orbital motion executed by the pair.

The results have been summarized in the tables. As expected, it is evident from the tables, that GW luminosities increase when one chooses larger values of masses while the electromagnetic charges decide the magnitude of EM wave power. From Table I and Table II, one notices a very interesting feature in that, when the BH masses are $\sim 10^{26}$ gm, electric charge for D2 is $q_2 \sim 10^{-5}\sqrt{G}\, m_2 - 10^{-3}\sqrt{G}\, m_2$ and magnetic charge for D1 is $g_1 = \hbar c/2q_2$, in consonant with Dirac quantization condition, the emanating EM power as well as the characteristic frequency are similar to what one observes for typical fast radio bursts (for a very short review on fast radio bursts (FRBs), see [89]). Of course, the standard explanations for FRBs hinge on astrophysical neutron stars endowed with very high magnetic fields $|\vec{B}| \gtrsim 10^{15}$ Gauss, lying amidst magnetized plasma.

Tables: Radiative Power and Characteristic Frequency - Electromagnetic and Gravitational Waves

Table 1. Average electromagnetic wave power, gravitational radiation power and characteristic frequency from a bound system, consisting of a magnetic BH and an electric BH with charge to mass ratio: $\frac{q_2}{\sqrt{G}m_2} = 1.0 \times 10^{-5}$ and magnetic charge: $g_1 = \frac{\hbar c}{2q_2}$; $R_{bh} \equiv 2Gm_2/c^2$.

Parameters						Results		
m_1	m_2	r_1/R_{bh}	r_2/R_{bh}	E (erg)	l (g cm$^2 s^{-1}$)	ω (rad s^{-1})	P_{em} (erg s^{-1})	P_{gw} (erg s^{-1})
29.0 $M_\odot$	36.0 $M_\odot$	20.6	617.4	-4.0×10^{52}	6.1×10^{52}	40.0	2.5×10^{40}	1.5×10^{50}
5.0×10^{26}g	5.0×10^{26}g	18.6	557.4	-3.9×10^{44}	3.4×10^{36}	7.0×10^{9}	5.7×10^{40}	4.2×10^{50}
2.9×10^{21}g	3.6×10^{22}g	34.4	1031.7	-1.2×10^{39}	2.6×10^{27}	2.8×10^{13}	3.2×10^{37}	6.8×10^{46}
1.0×10^{15}g	1.0×10^{16}g	33.8	1013.4	-4.3×10^{32}	2.4×10^{14}	1.0×10^{20}	5.2×10^{37}	1.2×10^{47}

Table 2. Average electromagnetic radiation power, gravitational wave power and characteristic frequency from a bound system consisting of a magnetic BH and an electric BH with charge to mass ratio: $\frac{q_2}{\sqrt{G}m_2} = 1.0 \times 10^{-3}$ and magnetic charge: $g_1 = \frac{\hbar c}{2q_2}$; $R_{bh} \equiv 2Gm_2/c^2$.

Parameters						Results		
m_1	m_2	r_1/R_{bh}	r_2/R_{bh}	E (erg)	l (g cm$^2 s^{-1}$)	ω (rad s^{-1})	P_{em} (erg s^{-1})	P_{gw} (erg s^{-1})
5.0×10^{26}g	6.0×10^{26}g	20.3	608.1	-3.5×10^{44}	4.4×10^{36}	5.0×10^{9}	2.8×10^{44}	1.7×10^{50}

Table 3. Average electromagnetic radiation power, gravitational wave power and characteristic frequency from a bound system consisting of a magnetic BH and an electric BH with charge to mass ratio: $\frac{q_2}{\sqrt{G}m_2} = 1.0 \times 10^{-3}$ and magnetic charge: $g_1 = \frac{\hbar c}{2e}$, where the electronic charge, $e = 4.8 \times 10^{-10}$ esu ; $R_{bh} \equiv 2Gm_2/c^2$.

Parameters						Results		
m_1	m_2	r_1/R_{bh}	r_2/R_{bh}	E (erg)	l (g cm$^2 s^{-1}$)	ω (rad s^{-1})	P_{em} (erg s^{-1})	P_{gw} (erg s^{-1})
10^{16}GeV$/c^2$	10^{20}g	38.0	682.0	-1.1×10^6	4.7×10^{-9}	5.0×10^{17}	1.0×10^6	9.1×10^{-9}

- **Set 1:** Parameters for the plots of Figure 2, Figure 3, Figure 4, Figure 5 and Figure 6:
  1. $g_1 = \sqrt{G}m_1 10^6, g_2 = 0, q_1 = 0, q_2 = \sqrt{G}m_2 10^{-4}, m_1 = 29M_\odot, m_2 = 36M_\odot, b \equiv Gm_1m_2 - (q_1q_2 + g_1g_2)$
  2. $R_{bh} = 2Gm_2/c^2$, $r_{\text{obs}} = 410$ Mpc, $r_2 = 617.43\ R_{bh}$, $r_1 = 20.6\ R_{bh}$
  3. $\phi_0 = \frac{\pi}{2}, l = 600\,\mu\, v\, R_{bh}$, $v = 0.01\, c$, $E = -\frac{\mu b^2}{16 l^2}$

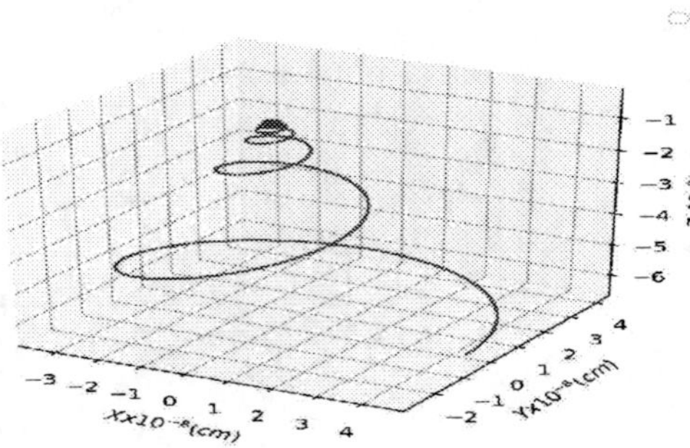

Figure 2. Set 1 parameters: Relative separation vector $\vec{r}$ between a gravitationally bound pair of electrically charged BH and a magnetically charged BH, starting from a maximum separation r_2, and coming closer to reach the closest distance of approach r_1. This plot represents the time evolution of the separation vector only for half the period of one cycle of the periodic motion.

- **Set 2:** Parameters for plots of Figure 7, Figure 8, Figure 9, Figure 10 and Figure 11:

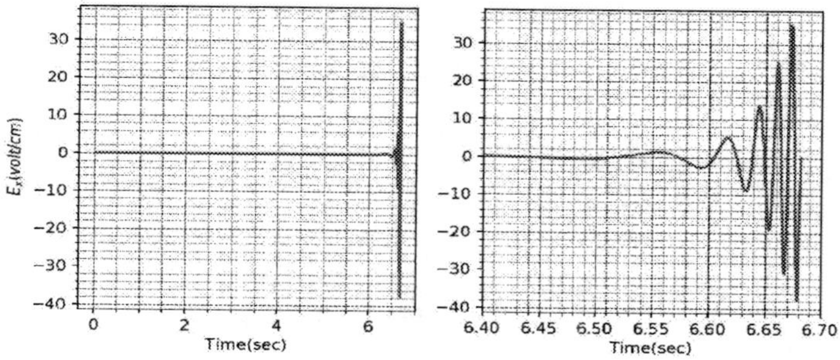

Figure 3. Set 1, EM wave amplitude: The x component of the electric field till the separation between the pair reaches the closest distance of approach r_1. The right hand side of the figure represents a zoomed version of the portion corresponding to the separations close to r_1.

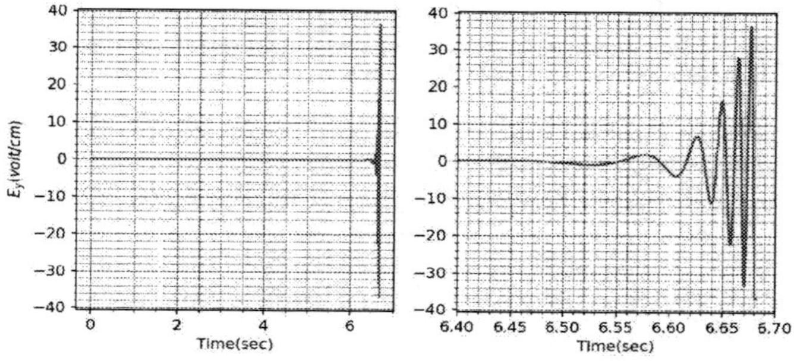

Figure 4. EM wave amplitude: The y component of the electric field till the separation between the pair reaches the closest distance of approach r_1. The right hand side of the figure represents a zoomed version of the portion corresponding to the separations close to r_1.

1. Electronic charge $e = 4.8 \times 10^{-10}$ e.s.u., $g_1 = \frac{hc}{4\pi e}, g_2 = 0, q_1 = 0, q_2 = \sqrt{G}m_2 10^{-5}, m_1 = 10^{16}$ GeV$/c^2, m_2 = 10^{20}$ gm, $b \equiv Gm_1m_2 - (q_1q_2 + g_1g_2)$
2. $R_{bh} = 2Gm_2/c^2, r_{\text{obs}} = 10^{13}$ cm, $r_2 = 682\ R_{bh}, r_1 = 38\ R_{bh}$.
3. $\phi_0 = \frac{\pi}{2}, l = 600\ \mu\, v\, R_{bh},\ v = 0.01\ c,\ E = -\ \frac{\mu b^2}{10 l^2}$.

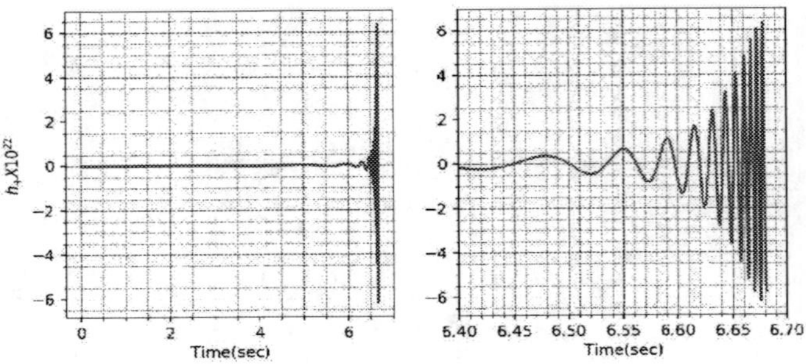

Figure 5. Set 1 GW amplitude: h_+ corresponding to the pair separation decreasing from r_2 to r_1.

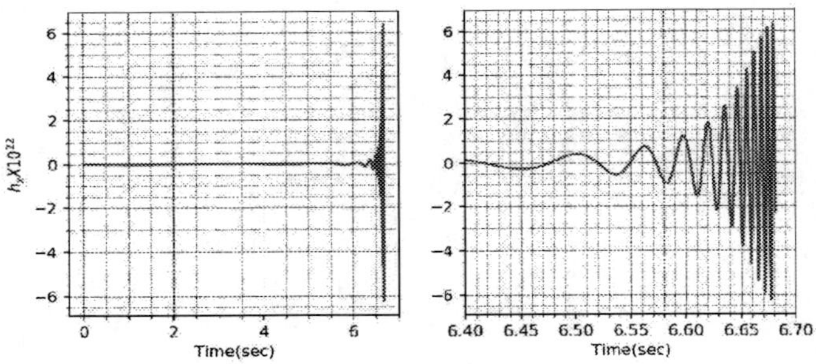

Figure 6. GW amplitude: $h_\times$ corresponding to the pair separation decreasing from r_2 to r_1.

Because of two distinct kinds of forces acting between the bound pair (electromagnetic and gravitational), there appears in the dynamics two different types of periodic motion. One is the fast gyration characterized by the instantaneous angular frequency ω, given by eqs. (36) and (71), arising due to the varying Lorentz force. The other is the non-planar periodic motion due to a combination of electrostatic and gravitational forces. Figure 2 and Figure 7 display the time evolution of the separation vector $\vec{r}$ between D1 and D2 from when they were maximally separated (r_2) to the closest distance of approach (r_1), corresponding to half of the slow periodic motion. The rapid gyration determined by ω is clearly reflected in the figures.

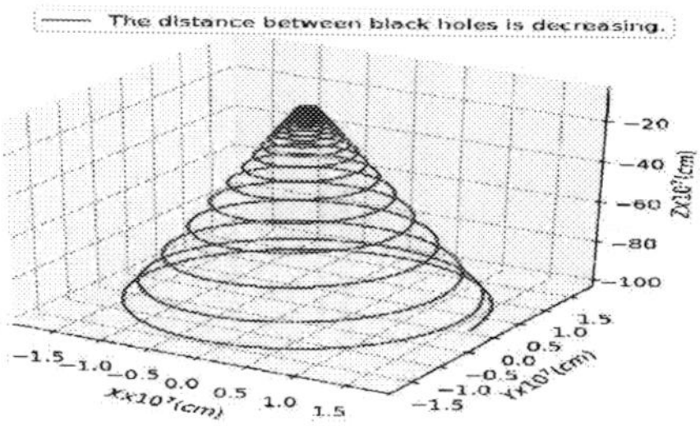

Figure 7. Set 2 parameters: Relative separation $\vec{r}$ between an electrically charged BH and a magnetically charged BH, bound gravitationally, starting from a maximum separation r_2, and coming closer to reach the closest distance of approach r_1, during half of its periodic motion.

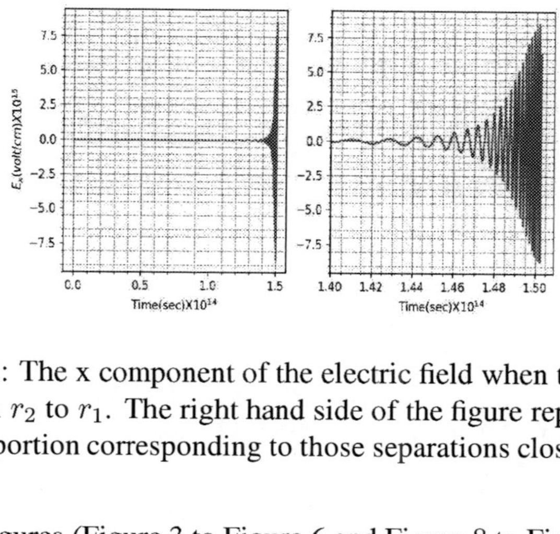

Figure 8. Set 2: The x component of the electric field when the pair separation decreases from r_2 to r_1. The right hand side of the figure represents a zoomed version of the portion corresponding to those separations close to r_1.

From the figures (Figure 3 to Figure 6 and Figure 8 to Figure 11), one finds that the wave amplitudes as well as the luminosities are very large when D1 and D2 approach r_1. This is expected since ω and the velocities increase sharply with decreasing separation r, entailing rapid changes in electromagnetic dipole moments as well as mass quadrupole moments. However, in the present analysis, we have not included the radiative reactions in our calculations. That will be the subject matter in an upcoming research paper of ours.

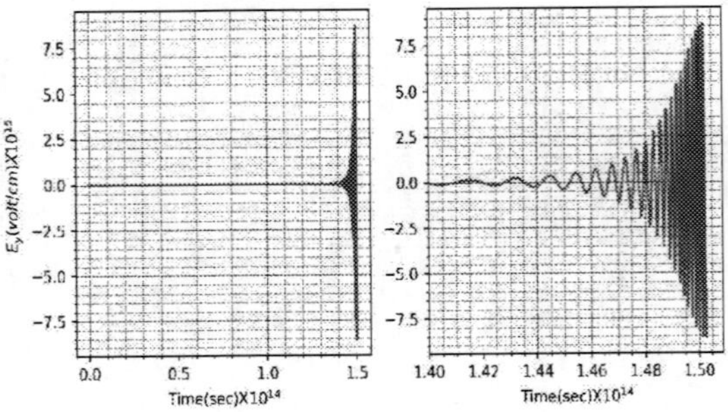

Figure 9. The corresponding y component of the electric field.

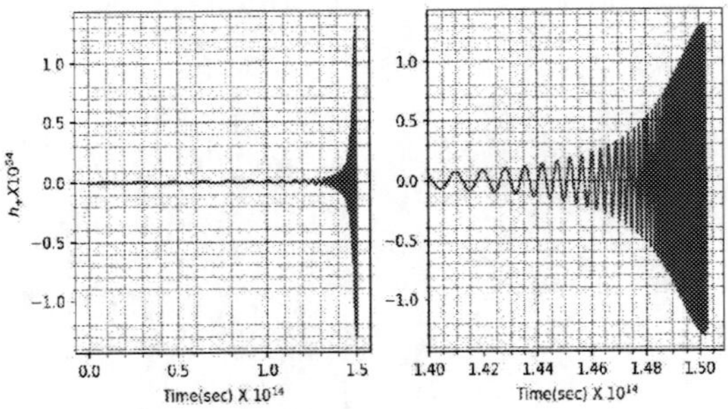

Figure 10. Set 2 GW amplitude: h_+ for the pair separation decreasing from r_2 to r_1.

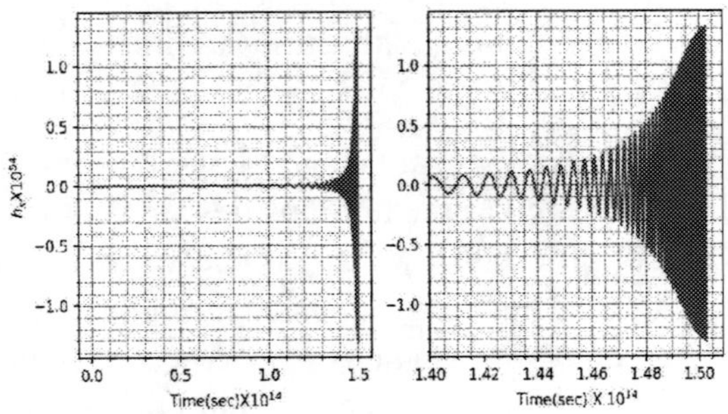

Figure 11. The corresponding behaviour of the $h_\times$ polarization.

Acknowledgments

PDG thanks Professor Debades Bandyopadhyay and Professor Debasish Majumdar, Saha Institute of Nuclear Physics, Kolkata, who had invited him to present the above research work of ours in the month of January, 2020, at the AAPCOS 2020 confererence. It is also a pleasure to thank Professor Ashoke Sen, Professor Samir Mathur, Professor Debasis Ghoshal and Professor Suvrat Raju for their valuable comments during several workshops and conferences.

References

[1] Penrose, R., 1965. Gravitational collapse and space-time singularities. *Physical Review Letters*, 14(3), p.57.

[2] Raychaudhuri, A., 1955. Relativistic cosmology. I. *Physical Review*, 98(4), p.1123.

[3] Vishveshwara, C. V., 1980. *Gravitation, Quanta and the Universe; proceedings of the Einstein Centenary Symposium held January 29 - February 3, 1979*, in Ahmedabad, India. Edited by A. R. Prasanna, J. V. Narlikar, and A Halsted Press Book, published by John Wiley & Sons, New York, 1980, p.154

[4] Srinivasan, G., *What are the Stars?* (University Press (India) Pvt. Ltd., Hyderabad, 2011)

[5] Srinivasan, G., *Can Stars Find Peace?* (University Press (India) Pvt. Ltd., Hyderabad, 2011)

[6] Das Gupta, P., 2012. S. Chandrasekhar: White Dwarfs, negative hydrogen ion,...., black holes, gravitational waves. *Physics Education*, **28**, (Jan-March 2012), p.5

(arxiv:1107.3460) [physics.hist-ph])

[7] Schwarzschild, K., *Über das Gravitationsfeld eines Massepunktes nach der Einsteinschen Theorie[*About the gravitational field of a mass point according to Einstein's theory*], Preussische Akademie der Wissenschaften, Sitzungberichte*, 1916, p. 189-196.

[8] Chandrasekhar, S., 1998. *The mathematical theory of black holes* (Vol. 69). Oxford university press.

[9] Calmet, X. and Hsu, S.D., 2022. Quantum hair and black hole information. *Physics Letters B*, 827, p.136995.

[10] Das Gupta, P., 2015. General Relativity and Compact Objects. RETCO-II: Theory and Observation, *ASI Conference Series*, Vol. 12, pp. 1 8. Edited by I. Chattopadhyay, A. Nandi, S. Das and S. Mandal.

[11] Volonteri, M., Habouzit, M. and Colpi, M., 2021. The origins of massive black holes. *Nature Reviews Physics*, 3(11), pp.732-743.

[12] Das Gupta, P. and Thareja, E., 2017. Supermassive black holes from collapsing dark matter BoseEinstein condensates. *Classical and Quantum Gravity*, 34(3), p.035006.

[13] Das Gupta, P. and Rahman, F., 2018. Aspects of Black Hole Physics and Formation of Super-massive Black Holes from Ultra-light Dark Bosons. *The Physical Universe*. Editors S. M. Wagh, S. D. Maharaj & G. Chon, 2018, ISBN Number: 978-93-5351-048-0. (arXiv:1801.02559 [gr-qc])

[14] Pereira, L.C., 2021. Progeny of supermassive black holes through the collapse of dark matter Bose-Einstein condensates: a variational approach. *Proceedings of the Romanian Academy Series A-Mathematics Physics Technical Sciences Information Science*.

[15] Hawking, S., 1971. Gravitationally collapsed objects of very low mass. *Monthly Notices of the Royal Astronomical Society*, 152(1), pp.75-78.

[16] Carr, B.J. and Hawking, S.W., 1974. Black holes in the early Universe. *Monthly Notices of the Royal Astronomical Society*, 168(2), pp.399-415.

[17] Zel'dovich, Y.B. and Novikov, I.D., 1967. The hypothesis of cores retarded during expansion and the hot cosmological model. *Soviet Astronomy*, 10, p.602.

[18] Das Gupta, P., 2012. General relativity and the accelerated expansion of the universe. *Resonance*, 17(3), pp.254-273.

[19] Carr, B., Kohri, K., Sendouda, Y. and Yokoyama, J.I., 2021. Constraints on primordial black holes. *Reports on Progress in Physics*, 84(11), p.116902.

[20] Khlopov, M.Y., Malomed, B.A. and Zeldovich, Y.B., 1985. Gravitational instability of scalar fields and formation of primordial black holes. *Monthly Notices of the Royal Astronomical Society*, 215(4), pp.575-589.

[21] Majumdar, A. S., Das Gupta, P. and Saxena, R.P., 1995. Baryogenesis from Black Hole Evaporation. *International Journal of Modern Physics D*, 4, pp. 517-529.

[22] Upadhyay, N., Das Gupta, P. and Saxena, R.P., 1999. Baryogenesis from primordial black holes after the electroweak phase transition. *Physical Review D*, 60(6), p.063513.

[23] Arun, K., Gudennavar, S.B. and Sivaram, C., 2017. Dark matter, dark energy, and alternate models: A review. *Advances in Space Research*, 60(1), pp.166-186.

[24] Hawking, S.W., 1974. Black hole explosions?. *Nature*, 248(5443), pp.30-31.

[25] Hawking, S.W., 1976. Black holes and thermodynamics. *Physical Review D*, 13(2), p.191.

[26] Page, D.N., 1993. Information in black hole radiation. *Physical review letters*, 71(23), p.3743.

[27] Mathur, S.D., 2002. A proposal to resolve the black hole information paradox. *International Journal of Modern Physics D*, 11(10), pp.1537-1540.

[28] Lunin, O. and Mathur, S.D., 2002. AdS/CFT duality and the black hole information paradox. *Nuclear Physics B*, 623(1-2), pp.342-394.

[29] Lunin, O. and Mathur, S.D., 2002. Statistical interpretation of the Bekenstein entropy for systems with a stretched horizon. *Physical review letters*, 88(21), p.211303.

[30] Mathur, S.D., 2005. The fuzzball proposal for black holes: An elementary review. *Fortschritte der Physik: Progress of Physics*, 53(78), pp.793-827.

[31] Mathur S.D. The information paradox: A pedagogical introduction *Classical Quantum Gravity*, 26 (2009), Article 224001

[32] Almheiri A., Marolf D., Polchinski J., Sully J. Black holes: Complementarity or firewalls? *J. High Energy Phys.*, 02 (2013), p. 062

[33] Raju, S., 2022. Lessons from the information paradox. *Physics Reports*, 943, pp.1-80.

[34] Denardo, G., Hively, L. and Ruffini, R., 1974. On the generalized ergosphere of the Kerr-Newman geometry. *Physics Letters B*, 50(2), pp.270-272.

[35] Penrose, R., 1969. Gravitational collapse: The role of general relativity. *Nuovo Cimento Rivista Serie*, 1, p.252.

[36] Penrose, R. and Floyd, R.M., 1971. Extraction of rotational energy from a black hole. *Nature Physical Science*, 229(6), pp.177-179.

[37] Bhat, M., Dhurandhar, S. and Dadhich, N., 1985. Energetics of the Kerr-Newman black hole by the Penrose process. *Journal of Astrophysics and Astronomy*, 6(2), pp.85-100.

[38] Wald, R.M., 1974. Black hole in a uniform magnetic field. *Physical Review D*, 10(6), p.1680.

[39] Lee, K., Nair, V.P. and Weinberg, E.J., 1992. A classical instability of Reissner-Nordstrm solutions and the fate of magnetically charged black holes. *Physical review letters*, 68(8), p.1100.

[40] Maldacena, J., 2021. Comments on magnetic black holes. *Journal of High Energy Physics*, 2021(4), pp.1-17.

[41] Ghosh, D., Thalapillil, A. and Ullah, F., 2021. Astrophysical hints for magnetic black holes. *Physical Review D*, 103(2), p.023006.

[42] Sivaram, C. and Sinha, K.P., 1977. Strong gravity, black holes, and hadrons. *Physical Review D*, 16(6), p.1975.

[43] Dirac, P.A.M., 1931. Quantised singularities in the electromagnetic field. *Proceedings of the Royal Society of London. Series A, Containing Papers of a Mathematical and Physical Character*, 133(821), pp.60-72.

[44] Saha, M.N., 1936. The origin of mass in neutrons and protons. *Indian Journal of Physics*, 10, 145.

[45] Preskill, J., 1984. *Magnetic monopoles. Annual Review of Nuclear and Particle Science*, 34(1), pp.461-530.

[46] Schwinger, J., 1969. A magnetic model of matter. *Science*, 165(3895), pp.757-761.

[47] Malik, R.P. and Pradhan, T., 1985. Local duality invariance of Maxwell equations. *Zeitschrift fr Physik C Particles and Fields*, 28(4), pp.525-527.

[48] Das Gupta, P., 1995. Magnetic charges and local duality symmetry in *Quantum Coherence and Reality*. Edited by J. S. Anandan & J. L. Safko, World Scientific, pp. 95-105.

[49] Das Gupta, P., 1993. Electric charge from spontaneous breaking of symmetry. *IUCAA Preprint*; 05/1993. (http://hdl.handle.net/11007/2791)

[50] Wilson, H.A., 1949. Note on Dirac's theory of magnetic poles. *Physical Review*, 75(2), p.309.

[51] Saha, M.N., 1949. Note on Dirac's theory of magnetic poles. *Physical Review*, 75(12), p.1968.

[52] Goldhaber, A.S., 1965. Role of spin in the monopole problem. *Physical Review*, 140(5B), p.B1407.

[53] Schwinger, J., 1975. Magnetic Charge and the Charge Quantization Condition, *Phys. Rev. D 12*, p. 3105.

[54] Wu, T.T. and Yang, C.N., 1976. Dirac monopole without strings: monopole harmonics. *Nuclear Physics B*, 107(3), pp.365-380.

[55] Lowe, D.A. and Strominger, A., 1994. Exact four-dimensional dyonic black holes and Bertotti-Robinson spacetimes in string theory. *Physical review letters*, 73(11), p.1468.

[56] Tseytlin, A.A., 1996. Extreme dyonic black holes in string theory. *Modern Physics Letters A*, 11(09), pp.689-713.

[57] Dijkgraaf, R., Verlinde, E. and Verlinde, H., 1997. Counting dyons in N= 4 string theory. *Nuclear Physics B*, 484(3), pp.543-561.

[58] Sen, A., 2007. Two centered black holes and N= 4 dyon spectrum. *Journal of High Energy Physics*, 2007(09), p.045.

[59] Sahoo, B. and Sen, A., 2007. α'-corrections to extremal dyonic black holes in heterotic string theory. *Journal of High Energy Physics*, 2007(01), p.010.

[60] Abishev, M.E., Boshkayev, K.A., Dzhunushaliev, V.D. and Ivashchuk, V.D., 2015. Dilatonic dyon black hole solutions. *Classical and Quantum Gravity*, 32(16), p.165010.

[61] 't Hooft, G., 1974. Magnetic monopoles in unified theories. *Nucl. Phys. B*, 79(CERN-TH-1876), pp.276-284.

[62] Polyakov, A.M., 1974. Pis ma Zh. Eksp. Teor. Fiz. 20, 430 (1974). *JETP Lett*, 20, p.194.

[63] Nambu, Y., 1974. Strings, monopoles, and gauge fields. *Physical Review D*, 10(12), p.4262.

[64] Julia, B. and Zee, A., 1975. Poles with both magnetic and electric charges in non-Abelian gauge theory. *Physical Review D*, 11(8), p.2227.

[65] Cho, Y.M. and Maison, D., 1997. Monopole configuration in Weinberg-Salam model. *Physics Letters B*, 391(3-4), pp.360-365.

[66] Cho, Y.M., Kimm, K. and Yoon, J.H., 2016. Gravitationally coupled electroweak monopole. *Physics Letters B*, 761, pp.203-206.

[67] Brihaye, Y., Hartmann, B. and Kunz, J., 1998. Gravitating dyons and dyonic black holes. *Physics Letters B*, 441(1-4), pp.77-82.

[68] Bjoraker, J. and Hosotani, Y., 2000. *Monopoles, Dyons and Black Holes in the Four-Dimensional Einstein-Yang-Mills Theory. Physical Review D*, 62(4), p.043513.

[69] Nolan, B.C. and Winstanley, E., 2012. On the existence of dyons and dyonic black holes in EinsteinYangMills theory. *Classical and Quantum Gravity*, 29(23), p.235024.

[70] Ellis, J., Mavromatos, N.E. and You,T., 2016. The Price of an Electroweak Monopole, *Phys. Lett. B*, 756, 29.

[71] Acharya, B., Alexandre, J., Baines, S., Benes, P., Bergmann, B., Bernabeu, J., Branzas, H., Campbell, M., Caramete, L., Cecchini, S. and De Montigny, M., 2017. Search for magnetic monopoles with the MoEDAL forward trapping detector in 13 TeV proton-proton collisions at the LHC. *Physical review letters*, 118(6), p.061801.

[72] Acharya, B., Alexandre, J., Baines, S., Benes, P., Bergmann, B., Bernabeu, J., Bevan, A., Branzas, H., Campbell, M., Caramete, L. and Cecchini, S., 2018. Search for magnetic monopoles with the MoEDAL forward trapping detector in 2.11 fb-inverse of 13 TeV protonproton collisions at the LHC. *Physics Letters B*, 782, pp.510-516.

[73] Cho, Y.M., Kimm, K., Oh, S. and Yoon, J.H., 2017. *Cosmological Production of Electroweak Monopole*. arXiv preprint arXiv:1703.02188.

[74] Burdin, S., Fairbairn, M., Mermod, P., Milstead, D., Pinfold, J., Sloan, T. and Taylor, W., 2015. Non-collider searches for stable massive particles. *Physics Reports*, 582, pp.1-52.

[75] Patrizii, L. and Spurio, M., 2015. Status of searches for magnetic monopoles. *Annual Review of Nuclear and Particle Science*, 65, pp.279-302.

[76] Zwanziger, D., 1968. Exactly soluble nonrelativistic model of particles with both electric and magnetic charges. *Physical review*, 176(5), p.1480.

[77] Das, S. and Majumdar, P., 1994. Electromagnetic charge-monopole versus gravitational scattering at Planckian energies. *Physical review letters*, 72(16), p.2524.

[78] Schwinger, J., Milton, K.A., Tsai, W.Y., DeRaad Jr, L.L. and Clark, D.C., 1976. Nonrelativistic DyonDyon scattering. In *A Quantum Legacy: Seminal Papers of Julian Schwinger* (pp. 550-594).

[79] Schwinger, J., 1966. Magnetic charge and quantum field theory. *Physical Review*, 144(4), p.1087.

[80] Zwanziger, D., 1968. Quantum field theory of particles with both electric and magnetic charges. *Physical Review*, 176(5), p.1489.

[81] Chandrasekhar, S., 1939. *An introduction to the study of stellar structure.* University of Chicago Press, Chicago.

[82] Damour, T. & Taylor, J. H., 1991. *Astrophys. J.*, 366, p. 501.

[83] Abbott, B.P., Abbott, R., Abbott, T.D., Abernathy, M.R., Acernese, F., Ackley, K., Adams, C., Adams, T., Addesso, P., Adhikari, R.X. and Adya, V.B., 2016. Observation of gravitational waves from a binary black hole merger. *Physical review letters*, 116(6), p.061102.

[84] Liebling, S.L. and Palenzuela, C., 2016. Electromagnetic luminosity of the coalescence of charged black hole binaries. *Physical Review D*, 94(6), p.064046.

[85] Landau, L.D. and Lifshitz, E.M., 1951. *The Classical Theory of Fields.* Addison-Wesley Press, Cambridge.

[86] Schutz, B. F., 2009. *A First Course in General Relativity.* Cambridge University Press, Cambridge.

[87] Maggiore, M. 2007. *Gravitational Waves Volume-1 Theory and Experiments*. Oxford University Press, Oxford.

[88] Peters, P.C. and Mathews, J., 1963. Gravitational radiation from point masses in a Keplerian orbit. *Physical Review*, 131, pp.435-440.

[89] Das Gupta, P. and Saini, N., 2018. Collapsing supra-massive magnetars: FRBs, the repeating FRB121102 and GRBs. *Journal of Astrophysics and Astronomy*, 39(1), pp.1-9.

Index

S

T

U

W

X

About the Editor

Dr. Kenath Arun
Assistant Professor
Department of Physics & Electronics
CHRIST (Deemed to be University)
Bengaluru, India
Email: kenath.arun@christuniversity.in